Skills Worksheet

Directed Reading A

Section: Natural Resources
EARTH'S RESOURCES

Write the letter of the correct answer in the space provided.

______ **1.** What is any natural material, such as air, that is used by humans called?
    a. natural resource
    b. alternative resource
    c. human resource
    d. conservation resource

______ **2.** Where does the energy from gasoline and wind come from?
    a. the oceans
    b. the sun
    c. the moon
    d. Earth's crust

Renewable Resources

______ **3.** What is a resource that can be replaced at the same rate at which it is used called?
    a. recyclable resource
    b. nonrenewable resource
    c. renewable resource
    d. Earth resource

______ **4.** Which of the following is a renewable resource?
    a. tree
    b. petroleum
    c. natural gas
    d. coal

Nonrenewable Resources

______ **5.** What is a resource that forms more slowly than the rate at which it is used called?
    a. renewable resource
    b. Earth resource
    c. recyclable resource
    d. nonrenewable resource

______ **6.** Which of the following is a nonrenewable resource?
    a. tree
    b. coal
    c. water
    d. air

Directed Reading A *continued*

CONSERVING NATURAL RESOURCES

_______ **7.** Which of the following is a way to conserve natural resources?
    a. Do not pollute lakes and rivers.
    b. Grow your own crops.
    c. Let water run when not in use.
    d. Save old magazines.

Energy Conservation: Reducing the Use

_______ **8.** Which of the following is a good way to conserve energy?
    a. Always use a car.
    b. Don't fill the washing machine.
    c. Leave the lights on.
    d. Take the bus.

Reusing and Recycling Resources

_______ **9.** What is reusing waste or scrap materials called?
    a. reducing
    b. refreshing
    c. recycling
    d. resourcing

_______ **10.** Which of the following is an advantage of recycling?
    a. Recycling increases the amount of trash.
    b. Recycling uses more energy than making new products.
    c. Recycling conserves energy.
    d. Recycling increases use of natural resources.

_______ **11.** Which of the following CANNOT be recycled?
    a. newspaper
    b. food
    c. aluminum can
    d. cardboard box

Directed Reading A

Section: Rock and Mineral Resources

MINERALS AND ROCKS

Write the letter of the correct answer in the space provided.

______ **1.** What is an inorganic solid that has a crystalline structure called?
- **a.** polymer
- **b.** mineral
- **c.** coal
- **d.** volcanic glass

______ **2.** Where do minerals form?
- **a.** in Earth's crust
- **b.** in Earth's core
- **c.** in chemicals
- **d.** in volcanic glass

______ **3.** Which of the following makes up most of the solid part of Earth's surface?
- **a.** silver
- **b.** gold
- **c.** rock
- **d.** mineral

Match the correct description with the correct term. Write the letter in the space provided.

______ **4.** Salt water dries up.

______ **5.** Pressure and temperature changes cause minerals to form.

______ **6.** Surface water carries dissolved materials into lakes.

- **a.** deposition
- **b.** evaporation
- **c.** metamorphism

Characteristics of Minerals

Write the letter of the correct answer in the space provided.

______ **7.** How many characteristics are used to identify a mineral?
- **a.** five
- **b.** four
- **c.** six
- **d.** two

Directed Reading A *continued*

_______ **8.** Why is coal not a mineral?
  a. because it is a synthetic material
  b. because it is black
  c. because it has crystals
  d. because it forms from the remains of plants

MINING

_______ **9.** Why are rocks and minerals mined from the ground?
  a. so we can build quarries
  b. because they cause pollution
  c. because there are too many
  d. so they can be made into objects

_______ **10.** What are deposits that are mined for profit called?
  a. glass
  b. surface minerals
  c. quarries
  d. ore

Surface Mining

_______ **11.** What is the purpose of surface mining?
  a. to dig tunnels
  b. to protect the habitats of plants
  c. to build shafts
  d. to remove ore from Earth's surface

_______ **12.** Which of the following is NOT a type of surface mine?
  a. shaft   **c.** open pit
  b. quarry   **d.** strip mine

Subsurface Mining

_______ **13.** What must be dug when ore is too deep within Earth?
  a. deposits
  b. tunnels
  c. open pits
  d. coal

Responsible Mining

_______ **14.** Which of the following is a problem that mining may create?
  a. reclamation
  b. mineral deposits
  c. energy conservation
  d. water pollution

Directed Reading A *continued*

_______ **15.** How do we return land used for mining to its original state?
   a. by pollution
   b. by recycling
   c. by reclamation
   d. by strip mining

MAKING COMMON OBJECTS

_______ **16.** Which of the following minerals can be used as it is?
   a. diamond
   b. bauxite
   c. gypsum
   d. calcite

Metals

_______ **17.** Which of the following is NOT true of metals?
   a. Metals let light pass through.
   b. Metals are good conductors of heat.
   c. Metals have shiny surfaces.
   d. Metals are good conductors of electricity.

_______ **18.** Which of the following is made with metals?
   a. plaster of Paris
   b. fertilizer
   c. aircraft
   d. cement

Nonmetals

_______ **19.** Which of the following is NOT true of nonmetals?
   a. Nonmetals may be translucent to light.
   b. Nonmetals are good conductors of electricity.
   c. Nonmetals are good insulators of heat.
   d. Nonmetals have shiny or dull surfaces.

_______ **20.** Which of the following is a nonmetallic mineral that helps make cement?
   a. diamond
   b. coal
   c. gypsum
   d. calcite

_______ **21.** Which of the following is used to make fertilizer?
   a. diamond
   b. gypsum
   c. calcite
   d. silica

Skills Worksheet

Directed Reading A

Section: Using Material Resources

Write the letter of the correct answer in the space provided.

______ **1.** Where do all of the objects you need to live come from?
   a. Earth
   b. atmosphere
   c. organisms
   d. sun

RESOURCES FROM EARTH

______ **2.** What kind of natural resources are used to generate energy?
   a. artificial resources
   b. energy resources
   c. material resources
   d. food resources

______ **3.** What kind of resources do humans use to make objects?
   a. material resources
   b. artificial resources
   c. atmospheric resources
   d. energy resources

Resources from the Atmosphere

______ **4.** Which of the following might be the most important resource in
   the atmosphere?
   a. iron
   b. argon
   c. nitrogen
   d. oxygen

______ **5.** How is argon used?
   a. as a fertilizer
   b. inside light bulbs
   c. to burn rocket fuel
   d. to make sand

Rock and Mineral Resources

______ **6.** What is iron used to make?
   a. polymers
   b. argon
   c. steel
   d. copper wiring

_______ **7.** Where is salt harvested from?
    a. the atmosphere
    b. organisms
    c. sea water
    d. lakes

Petroleum

Use the terms from the following list to complete the sentences below.

   plastics   petroleum

8. A liquid mixture of hydrocarbons used as a fuel is called

_______________________.

9. Polymers, which are made from chemicals, are also called

_______________________.

RESOURCES FROM LIVING THINGS
Plant Resources

Match the correct description with the correct term. Write the letter in the space provided.

_______ **10.** a source of energy for humans   **a.** fruits and nuts

_______ **11.** a source of lumber and paper   **b.** cotton

_______ **12.** a source of clothing   **c.** trees

Animal Resources

Write the letter of the correct answer in the space provided.

_______ **13.** Animals are NOT used to make which of the following products?
    a. clothing
    b. food
    c. leather
    d. cotton

_______ **14.** Which of the following animal resources can be used to fertilize crops?
    a. wastes
    b. fibers
    c. dairy products
    d. leather

Directed Reading A *continued*

THE COSTS OF MATERIAL RESOURCES
Economic Costs

_______ **15.** What is the total cost of making a product called?
- **a.** pollution cost
- **b.** natural cost
- **c.** environmental cost
- **d.** economic cost

Environmental Costs

_______ **16.** What is the damage to the land caused by making a product called?
- **a.** pollution cost
- **b.** natural cost
- **c.** environmental cost
- **d.** economic cost

_______ **17.** Which of the following may help manufacturers lower an environmental cost?
- **a.** pollution control
- **b.** marketing
- **c.** transport
- **d.** feeding

Using Resources Wisely

_______ **18.** What is one way to lower the economic cost of using a natural resource?
- **a.** controlling pollution
- **b.** improving manufacturing
- **c.** having faster transportation
- **d.** using more resources

_______ **19.** What is one way to lower the environmental cost of using a natural resource?
- **a.** investing in marketing
- **b.** using cheaper resources
- **c.** restoring habitats
- **d.** using more resources

Skills Worksheet)

Directed Reading B

Section: Natural Resources

1. What do the water you drink, the paper you write on, and the air you breathe have in common?

____

____

EARTH'S RESOURCES

______ **2.** Any natural material that is used by humans is called a(n)
- **a.** human resource.
- **b.** Earth resource.
- **c.** natural resource.
- **d.** recyclable resource.

______ **3.** Which of the following is NOT a natural resource?
- **a.** mineral
- **b.** petroleum
- **c.** forest
- **d.** plastic

4. Where does the energy we get from many of our natural resources ultimately come from?

____

5. What is a renewable resource?

____

____

____

6. Give one example of a renewable resource.

____

7. What is a nonrenewable resource?

____

____

____

8. Give three examples of a nonrenewable resource.

____

____

____

Directed Reading B *continued*

CONSERVING NATURAL RESOURCES

9. What is one way in which people can conserve natural resources?

____

____

____

10. What is one reason lakes, rivers, and other water resources should be kept free of pollution?

____

____

11. Name two ways in which people can conserve energy.

____

____

12. Define *recycling.*

____

____

____

13. How does recycling help conserve energy?

____

____

____

____

14. Name three kinds of products that are recyclable.

____

____

____

Name _________________________________ Class _______________ Date ___________

Directed Reading B

Section: Rock and Mineral Resources

1. A naturally formed solid that has a crystalline structure is a(n)

_______________________________.

2. Where do minerals form?

____

____

3. The natural material that makes up most of the solid part of Earth is

_______________________________.

Match the correct description with the correct term. Write the letter in the space provided.

_______ **4.** A body of water dries up.

_______ **5.** Magma rises upward and cools.

_______ **6.** Groundwater is heated by magma.

_______ **7.** Surface water and groundwater carry
dissolved materials into lakes and seas.

_______ **8.** Changes take place in pressure,
temperature, and chemical makeup.

a. metamorphism

b. cooling of plutons

c. deposition

d. reactions between
hot water and rocks

e. evaporation

9. Name the five characteristics geologists use to identify a mineral.

____

____

____

____

____

____

Directed Reading B *continued*

MINING

10. Why are rocks and minerals mined from the ground?

____

____

11. A deposit of desired material large enough to be mined for profit is called

a(n) _____________________.

12. Name the two methods by which rocks and minerals are removed from the ground.

____

____

13. What kind of mines are open pits and quarries?

____

14. When coal is removed in strips, it is called _____________________.

15. What is used when ore is too deep within Earth to be surface mined?

____

16. What is dug into the ground in subsurface mining?

____

17. List two problems that can result from mining.

____

____

18. The process of returning land to its original state after mining is called

_____________________.

MAKING COMMON OBJECTS

19. Name a mineral that can be used as it is.

____

20. Name three characteristics of metals.

____

____

____

Directed Reading B *continued*

21. Name three characteristics of nonmetals.

____

____

____

22. What is a major component of cement?

____

23. What is gypsum used to make?

____

____

____

Directed Reading B

Section: Using Material Resources

1. Where do all of the objects that you need to live come from?

____

RESOURCES FROM EARTH

_______ **2.** Into what two groups can Earth's resources be divided?
   a. natural and artificial
   b. energy and material
   c. natural and material
   d. energy and organic

_______ **3.** What kind of resources do humans consume or use to make objects?
   a. material
   b. natural
   c. atmospheric
   d. energy

4. List the four sources on Earth for material resources.

____

____

____

____

5. What is the most valuable resource in the atmosphere?

____

Match the correct description with the correct term. Write the letter in the space provided.

_______ **6.** used as a source of chlorine

_______ **7.** used to help reduce car exhaust

_______ **8.** used inside light bulbs

_______ **9.** used to provide gasoline

_______ **10.** used to make steel

_______ **11.** used to make pipes for plumbing

a. petroleum
b. iron
c. polymer
d. argon
e. platinum
f. salt

Name _______________________________ Class _______________ Date _______________

12. A liquid mixture of complex hydrocarbons is called

_______________________.

RESOURCES FROM LIVING THINGS

13. Why do humans harvest and consume plants?

____

____

14. Name three products that people use that are made from plant resources.

____

____

____

15. Name five products that people use that are made from animal resources.

____

____

____

____

____

THE COSTS OF MATERIAL RESOURCES

16. Name two examples of costs related to using plant and animal resources.

____

____

17. What is the basic requirement for commercial products?

____

____

18. What must be used if a material resource becomes too expensive to obtain?

____

____

Directed Reading B *continued*

19. Name two possible environmental costs of a product.

____

____

20. What are three things that some environmental protection laws require of a community?

____

____

____

Skills Worksheet

Vocabulary and Section Summary A

Natural Resources

VOCABULARY

In your own words, write a definition of the following terms in the space provided.

1. natural resource

____

____

2. renewable resource

____

____

3. nonrenewable resource

____

____

4. recycling

____

____

SECTION SUMMARY

Read the following section summary.

- We use natural resources such as fresh water, petroleum, and trees to make our lives easier and more comfortable.

- Renewable resources can be replaced in a relatively short time, but nonrenewable resources may take thousands or even millions of years to form.

- Natural resources can be conserved by using only what is needed, by taking care of resources, and by reusing and recycling.

Vocabulary and Section Summary A

Rock and Mineral Resources

VOCABULARY

In your own words, write a definition of the following terms in the space provided.

1. mineral

____

____

____

2. ore

____

____

____

SECTION SUMMARY

Read the following section summary.

- A mineral is a naturally formed, inorganic solid that has a definite crystalline structure and a consistent chemical composition.

- Environments in which minerals form may be located at or near Earth's surface or deep below the surface.

- Two types of mining are surface mining and subsurface mining.

- Two ways to reduce the harmful effects of mining are through the reclamation of mined land and the recycling of mineral products.

- Both metals and nonmetals are used to make common objects.

Vocabulary and Section Summary A

Using Material Resources

VOCABULARY

In your own words, write a definition of the following terms in the space provided.

1. material resource

____

____

____

2. petroleum

____

____

____

SECTION SUMMARY

Read the following section summary.

- Resources from Earth include gases from the atmosphere and rocks, minerals, and petroleum from Earth's crust.

- Living things provide humans with materials, such as food, clothing, and shelter.

- Using natural resources involves both economic and environmental costs.

- Reducing the environmental cost of using resources sometimes involves increasing the economic cost.

Vocabulary and Section Summary B

Natural Resources

VOCABULARY

After you finish reading the section, try this puzzle! In each statement below, the underlined term is incorrect. Write the correct term in the space provided.

1. Any natural material that is used by humans is a(n) <u>atmospheric resource</u>.

____

2. <u>Mining</u> is the process of reusing materials from waste or scrap.

____

3. An example of a(n) <u>renewable resource</u> is coal, which takes millions of years to form.

____

4. An example of a(n) <u>environmental resource</u> is fresh water, which can be replaced in a relatively short time.

____

5. Earth's atmosphere maintains air <u>composition</u> and produces rain.

____

SECTION SUMMARY

Read the following section summary.

- We use natural resources such as fresh water, petroleum, and trees to make our lives easier and more comfortable.

- Renewable resources can be replaced in a relatively short time, but nonrenewable resources may take thousands or even millions of years to form.

- Natural resources can be conserved by using only what is needed, by taking care of resources, and by reusing and recycling.

Skills Worksheet

Vocabulary and Section Summary B

Rock and Mineral Resources

VOCABULARY

After you finish reading the section, try this puzzle! Use the clues below to unscramble the letters, and write the correct terms in the blanks. Then, write the corresponding letters in the numbered blanks to answer the question.

1. a natural material that makes up most of the solid part of Earth's surface: CKRO

2. a process by which land used for mining is returned to its original state: LMAATRIENOC

3. a natural, inorganic solid that has an orderly internal structure: EALNIMR

4. what surface coal mining is sometimes known as: PTSIR GNIIMN

5. a natural material whose concentration of economically valuable minerals is high enough to be mined profitably: REO

6. a process during which there are changes in pressure, temperature, or chemical makeup: MOPMSMEATRHI

1. ____ ____ ____ ____
     5   9

2. ____ ____ ____ ____ ____ ____ ____ ____ ____ ____ ____
      1   4

3. ____ ____ ____ ____ ____ ____ ____
 3

4. ____ ____ ____ ____ ____ ____ ____ ____ ____ ____ ____
    10   6

5. ____ ____ ____
  2   8

6. ____ ____ ____ ____ ____ ____ ____ ____ ____ ____ ____
         7   11

7. What do minerals and ores make?

____ ____ ____ ____ ____ ____
 1   2   3   4   5   6

____ b j ____ ____ ____ ____
 7   8   9 10 11

Vocabulary and Section Summary B *continued*

SECTION SUMMARY

Read the following section summary.

- A mineral is a naturally formed, inorganic solid that has a definite crystalline structure and a consistent chemical composition.

- Environments in which minerals form may be located at or near Earth's surface or deep below the surface.

- Two types of mining are surface mining and subsurface mining.

- Two ways to reduce the harmful effects of mining are through the reclamation of mined land and the recycling of mineral products.

- Both metals and nonmetals are used to make common objects.

Vocabulary and Section Summary B

Using Material Resources

VOCABULARY

After you finish reading the section, try this puzzle! Use the clues below to solve the crossword puzzle.

ACROSS

2. polymers made from chemicals

3. a natural resource that humans use to make objects

4. the process of recovering useful materials from waste

DOWN

1. a natural, usually inorganic solid that has a characteristic chemical composition

2. a resource used to provide gas

SECTION SUMMARY

Read the following section summary.

- Resources from Earth include gases from the atmosphere and rocks, minerals, and petroleum from Earth's crust.

- Living things provide humans with materials, such as food, clothing, and shelter.

- Using natural resources involves both economic and environmental costs.

- Reducing the environmental cost of using resources sometimes involves increasing the economic cost.

Reinforcement

What Are My Resources?

Complete this worksheet after you finish reading the section "Natural Resources."

Something that people use that comes from Earth is known as a natural resource. There are a lot of natural resources on Earth, and they are broken up into two types—renewable and nonrenewable. You might be wondering what the difference is between these two. Renewable resources, such as trees, can be replaced in a relatively short time after they are used. But a nonrenewable resource, such as coal, can take thousands or millions of years to replace. Because it takes such a long time to replace nonrenewable resources, whatever amount exists on Earth right now is limited.

Take a look at the pictures below and label each item with an *R* if it is renewable or with an *N* if it is nonrenewable. Write your answer in the space provided.

Critical Thinking

Material Morning

It's Tuesday, 7:00 a.m. Sally gets out bed and feels the softness of the cotton rug on the floor at her feet. She likes coffee in the morning, so she goes into the kitchen, turns on the light, and makes a full pot of coffee. While she sips her coffee, she reads the newspaper. After a quick breakfast of eggs and toast, Sally takes a long, hot shower, brushes her teeth, and gets dressed. Then, she hops into her car and heads to work.

COMPREHENDING IDEAS

1. Can you name all of the products that come from natural resources that Sally came in contact with in just one morning?

____

____

____

____

____

2. Does the newspaper Sally read come from a renewable resource? Explain your answer.

____

____

____

____

3. What material resources went into making the bread that Sally toasted for breakfast?

____

____

____

____

Critical Thinking *continued*

FORMING AN OPINION

4. Would the demand for these natural resources be greater in an apartment
building or a single-family house? Explain your answer.

____

____

____

____

DEMONSTRATING REASONED JUDGMENT

5. How could Sally conserve the resources that she uses every morning?

____

____

____

____

Activity

SciLinks Activity

RECYCLING

Go to www.scilinks.org. To find links related to recycling, type in the keyword HY71277. Then, use the links to do the following recycling activity.

Internet Resources

For a variety of links related to this chapter, go to www.scilinks.org

Topic: Recycling
SciLinks code: HY71277

Imagine that you are an object in a recycling bin, such as an old newspaper, an aluminum soda can, a glass pickle jar, a plastic grocery bag, or a plastic toy. Use the space below to write a creative story about your journey from the bin to a new life. Include descriptions of the recycling steps you go through and what new recycled product you become. Give some advantages of recycling the object you choose.

____

____

____

____

____

____

____

____

____

____

____

____

____

____

Skills Worksheet

Section Review

Natural Resources

USING VOCABULARY

1. Write an original definition for *natural resource, recycling, renewable resource,* and *nonrenewable resource.*

UNDERSTANDING CONCEPTS

2. Summarizing How do humans use most natural resources?

3. Comparing Compare the rates at which renewable and nonrenewable resources form.

CRITICAL THINKING

4. Applying Concepts Describe three ways you could conserve natural resources.

Section Review *continued*

5. Making Inferences How can people's actions affect renewable and nonrenewable resources?

____

____

____

MATH SKILLS

6. Making Calculations A faucet drips for 8.6 h, and 3.3 L of water drips out every hour. What is the total number of liters of water that drip out? Show your work below.

CHALLENGE

7. Predicting Consequences An island mostly covered with trees was home to a small group of humans. The trees were a renewable resource. How might a change in the human population affect whether the trees remain a renewable resource?

____

____

____

____

____

____

____

Skills Worksheet

Section Review

Rock and Mineral Resources

USING VOCABULARY

1. Write an original definition for *mineral* and *ore*.

UNDERSTANDING CONCEPTS

2. Identifying What are five things that must be true in order for a substance to be considered a mineral?

3. Describing Are minerals renewable resources or nonrenewable resources?

4. Comparing What are the two main types of mining, and how do they differ?

5. Listing List three metals, and name a common object made from each metal.

CRITICAL THINKING

6. Making Inferences Name two minerals that might be found at the site of an ancient ocean.

Section Review *continued*

7. Analyzing Ideas How does reclamation protect the environment around a mine?

____

____

____

8. Applying Concepts How does the recycling of mineral products reduce the costs of manufacturing objects from minerals?

____

____

____

MATH SKILLS

9. Making Calculations A copper mine has removed 120,000 metric tons of earth from the mine in a week. If the copper ore at the mine contains 0.7% copper, how much copper was mined during this time? Show your work below.

CHALLENGE

10. Making Inferences A variety of economic and technological factors determine whether a rock or mineral deposit is considered an ore. Name two factors that could influence whether a deposit is considered an ore. Explain your answers.

____

____

____

____

____

____

Section Review

Using Material Resources

USING VOCABULARY

1. Write an original definition for *material resource* and *petroleum*.

UNDERSTANDING CONCEPTS

2. Listing What are three resources that can be obtained from the atmosphere? How are these resources used?

3. Describing Explain how petroleum is used to make common objects.

4. Identifying Give three examples of common objects that are made from plant resources.

CRITICAL THINKING

5. Analyzing Relationships Why does protecting the environment sometimes result in an increased price in the store for a product?

| Section Review *continued*

---

6. Expressing Opinions Explain why you think it is or is not important to consider environmental costs when determining the price of manufactured materials.

____

____

____

7. Applying Concepts List five material resources. List whether they are renewable or nonrenewable. Explain your answers.

____

____

____

____

____

MATH SKILLS

8. Making Calculations In 1990, world rice production was 350 million tons. In 1999, it had risen to 400 million tons. What was the percentage increase in world rice production between 1990 and 1999? Show your work below.

CHALLENGE

9. Analyzing Processes Which energy and material resources are used to make new paper? Which energy and material resources are used to make paper from recycled fiber? Do you think it makes sense to recycle paper? Explain your answer.

____

____

____

____

Chapter Review

USING VOCABULARY

1. Academic Vocabulary In the sentence "A mining company will consider the distribution of coal when deciding how to mine the coal," what does the word *distribution* mean?

____

____

2. Write an original definition for *natural resource* and *material resource*.

____

____

For each pair of terms, explain how the meanings of the terms differ.

3. *mineral* and *ore*

____

____

4. *renewable resource* and *nonrenewable resource*

____

____

UNDERSTANDING CONCEPTS
Multiple Choice

_______ **5.** Which of the following resources is a renewable resource?
- **a.** coal
- **b.** trees
- **c.** petroleum
- **d.** iron ore

_______ **6.** Which of the following resources is a nonrenewable resource?
- **a.** copper ore
- **b.** trees
- **c.** fresh water
- **d.** wildlife

Chapter Review *continued*

_______ **7.** The process by which land used for mining is returned to its original
state is called
- **a.** recycling.
- **b.** regeneration.
- **c.** reclamation.
- **d.** renovation.

_______ **8.** Natural resources can be conserved by
- **a.** using only what is needed.
- **b.** protecting the quality of resources such as air and water.
- **c.** reusing and recycling.
- **d.** All of the above

_______ **9.** Which of the following can be made from metals?
- **a.** plaster board
- **b.** bicycle frames
- **c.** glass
- **d.** All of the above

_______ **10.** Which of the following can be made from trees?
- **a.** rubber
- **b.** paper
- **c.** lumber
- **d.** All of the above

Short Answer

11. Listing Describe three ways that humans use natural resources.

____

____

____

12. Summarizing Explain the five characteristics of a mineral.

____

____

____

____

____

13. Describing Describe two environments in which minerals form. List a mineral
that is formed in each environment.

____

____

____

| Chapter Review *continued*

---

14. Listing List two common objects for which the natural origin is a metal and two common objects for which the natural origin is a nonmetal.

____

____

15. Listing List two common objects for which the natural origin is a plant resource and two common objects for which the natural origin is an animal resource.

____

____

16. Comparing Compare surface and subsurface mining.

____

____

17. Identifying Identify two common objects that are made from petroleum.

____

____

WRITING SKILLS

18. Outlining Topics Outline the process of making paper from trees.

____

____

____

____

____

____

____

____

____

____

Chapter Review *continued*

CRITICAL THINKING

19. Concept Mapping Use the following terms to create a concept map: *natural resources, renewable resources, nonrenewable resources, material resources, petroleum, iron, trees,* and *energy resources.*

20. Evaluating Assumptions Why do we need to conserve renewable resources even though they can be replaced?

____

____

____

| Chapter Review *continued*

21. Applying Concepts Describe the different ways that you can conserve natural resources at home.

____

____

____

22. Applying Concepts Could a resource that was once renewable become nonrenewable? Explain your answer.

____

____

____

INTERPRETING GRAPHICS

Use the graph below to answer the next three questions.

**Average Lifetime Mineral and Metal
Consumption per Person in the United States**

23. Evaluating Data Which resource is used the most? Which resource is used the least?

____

____

24. Applying Concepts Are the resources shown in the graph energy resources or material resources? Explain your answer.

____

____

____

Chapter Review *continued*

---

25. Making Inferences Explain how the practice of recycling could affect the information on the graph for aluminum and copper.

____

____

____

26. Making Comparisons Explain the difference between metals and nonmetals. Give examples of each.

____

____

____

____

____

27. Applying Concepts Give examples of environmental concerns that would be taken into account by a mining company as it creates a reclamation plan for an open-pit mine.

____

____

____

____

____

28. Identifying Relationships Explain why the environmental costs of producing paper may or may not be reflected in the price of the paper in the store.

____

____

____

____

____

| Chapter Review *continued*

---

29. Predicting Consequences Imagine that a nonrenewable material resource used to make computers begins to run out. Devise two different plans to address the shortage. Explain how each plan would work and how each plan might affect the price of computers.

____

____

____

____

____

____

MATH SKILLS

Interpreting Graphics

Use the table below to answer the next three questions.

Mass of Paper Products		
Product	**Tons produced**	**Percentage recycled**
Newspapers	13,620	56.4
Books	1,140	14.0
Magazines	2,260	20.8
Office papers	7,040	50.4

30. Making Calculations How many tons of all types of paper products were generated? Show your work below.

31. Making Calculations How many tons of newspapers were recycled? Show your work below.

32. Making Calculations How many total tons of paper products in all categories were recycled? Show your work below.

CHALLENGE

33. Analyzing Relationships Petroleum is used to generate energy and to make common objects such as plastics. Describe how the uses of petroleum might change as petroleum becomes more scarce.

____

____

____

____

____

____

____

____

____

____

____

____

____

____

Chapter Pretest

Teacher Notes and Answer Key

The Pretest questions are designed to help you determine the prior knowledge of your students. Some questions test whether students have mastered the background knowledge they need to understand the content you are about to teach. Other questions test your students' prior knowledge of the content you are about to teach. Use the Pretest with the Test Doctors and diagnostic teaching tips in these notes pages to help you tailor your instruction to your students' specific needs.

QUESTION NUMBER	CORRECT ANSWER	STANDARD
1	A	5.3.a
2	B	5.3.c
3	D	5.1.c
4	D	6.5.b
5	C	6.5.a
6	C	6.6.b
7	D	6.6.c
8	A	6.6.b
9	B	6.6.c
10	B	6.6.b

TEST DOCTOR

The following Pretest questions have been diagnosed by the Test Doctor. Find out what might be causing your students' "ailing" answers. Each Test Doctor is followed by a diagnostic teaching tip to help you address students' learning needs.

Question 1 *asks students to identify where most of Earth's water is located.*

A Correct. About 96% of Earth's water is in oceans, seas, and bays.

B Incorrect. About 0.75% of Earth's water is under the ground in the form of fresh water.

C Incorrect. About 0.007% of Earth's water is in rivers and lakes in the form of fresh water.

D Incorrect. About 1.7% of Earth's water is in glaciers, ice caps, and permanent snow.

Diagnostic Teaching Tip: Students who have difficulty with this question should benefit from a review of water resources on Earth. Obtain a list of water resources, such as the one from the U.S. Department of the Interior and the U.S. Geological Survey. Have students review the information. You may make a math connection with this activity if you choose to have the class work together to create a pie graph from the data they find.

Question 2 *asks students to identify sources of fresh water.*

A Incorrect. Private wells are often sources of drinking water in California.

B Correct. The San Francisco Bay contains salt water and is not a source of drinking water.

| Chapter Pretest *continued*

C Incorrect. The San Joaquin River is a preferred source of drinking water in California.
D Incorrect. Lake Tahoe is an important freshwater lake that tourists visit for water activities.

Diagnostic Teaching Tip: Students who answer this question incorrectly should benefit from a review of sources of fresh water and salt water. Begin by telling students that although fresh water is a renewable resource, it is important to conserve it. Remind students that oceans, seas, and bays are sources of salt water. Most other sources of water are sources of fresh water. Students who answer this question incorrectly probably do not realize that the San Francisco Bay is a source of salt water.

Question 3 *asks students to know that metals are good conductors of electricity.*

A Incorrect. Plants are not good conductors of electricity, and they cannot be made into wire. The cacao tree's seeds are used to make cocoa and chocolate.
B Incorrect. Nonmetals are not good conductors of electricity, and they cannot be made into wire. Carbon is present in minerals and in all living organisms.
C Incorrect. Minerals cannot be made into wire. Clay is often used to make ceramics and bricks.
D Correct. Copper is a metal, so it is a good conductor of electricity and can be formed into wires. Copper is used in electrical wiring.

Diagnostic Teaching Tip: Students who have difficulty answering this question correctly might benefit from a step-by-step analysis of each of the responses. Although water-saturated clay conducts electrical current, clay cannot be formed into wire. Plants and nonmetals are not good conductors, and they cannot be made into wire. Remind students that metals are known to be good conductors of electricity, and they can be made into wire. Tell students that they will be studying how different types of natural resources are used to make things in the everyday world.

Question 4 *asks students to identify processes in the rock cycle.*

A Incorrect. Cementation changes sediment to sedimentary rock.
B Incorrect. Pressure changes sediment to sedimentary rock, sedimentary rock to metamorphic rock, and igneous rock to metamorphic rock.
C Incorrect. Cooling changes magma to igneous rock.
D Correct. Evaporation is found in the water cycle, not the rock cycle.

Diagnostic Teaching Tip: Students who have difficulty answering this question correctly should benefit from a review of the rock cycle, as found in Section 4, "The Cycling of Matter," in Chapter 3, "Earth's Systems and Cycles." Have volunteers trace the various pathways and tell the processes that make each type of rock. Tell students that they will be studying how different types of rocks and minerals are used to make things in the everyday world.

Chapter Pretest *continued*

Question 5 *asks students how energy is transferred through an ecosystem.*

A Incorrect. Energy is not transferred to plants when animals eat them.

B Incorrect. Sunlight is not kinetic energy.

C Correct. Solar energy is converted into chemical energy through photosynthesis, and then transferred to animals when they eat plants.

D Incorrect. Sunlight is not mechanical energy. Energy is not transferred to plants when animals eat them.

Diagnostic Teaching Tip: Students who have difficulty answering this question correctly should benefit from a review of the carbon cycle, as found in Section 4, "The Cycling of Matter," in Chapter 3, "Earth's Systems and Cycles." Ask students to describe how sunlight provides energy for plants and animals on Earth.

Question 6 *asks students to identify the characteristics of a mineral.*

A Incorrect. Magma is molten rock. It is not solid with an orderly structure.

B Incorrect. Ore is a large rock or mineral deposit.

C Correct. Minerals are naturally formed. They are solids and usually form by inorganic processes. They have a uniform chemical composition and are crystals.

D Incorrect. Rock can be completely or partially made of mineral.

Diagnostic Teaching Tip: Students who have difficulty answering this question correctly might benefit from an introduction of rock and mineral resources. Have small groups of students research minerals in an encyclopedia or on the Internet. Have groups list the characteristics that define a mineral and give five examples of minerals. Be sure to emphasize Section 2, "Rock and Mineral Resources," in Chapter 4, "Material Resources." Students must be able to identify different material resources, including rocks and minerals, in order to master standard 6.6.b.

Question 7 *asks students what common mineral is used in the production of computer chips.*

A Incorrect. Copper is used in electrical wire and plumbing.

B Incorrect. Diamond is used in jewelry, and cutting and drilling tools.

C Incorrect. Halite is used in highway de-icers and water softeners.

D Correct. Silicon in quartz is used to make computer chips.

Diagnostic Teaching Tip: Students who have difficulty answering this question correctly might benefit from an introduction to one use of rocks and minerals—making computer chips. Have small groups research on the Internet how silicon computer chips are made. Ask each group to prepare an advertisement for a new computer chip, incorporating the information they researched on how it was made. Be sure to emphasize Section 2, "Rock and Mineral Resources," in Chapter 4, "Material Resources." Students must know the natural origin of the materials used to make common objects in order to master standard 6.6.c.

Question 8 *asks students to identify sources of material resources.*

A Correct. The sun is a source of energy, but it is not a source of material resources or natural resources that humans consume or use to make objects.

B Incorrect. The atmosphere is a source of oxygen, nitrogen, and other gases.

C Incorrect. The ocean holds many resources, such as organisms for food.

D Incorrect. Plants and animals are material resources for food and clothing.

Diagnostic Teaching Tip: Students who have difficulty with this question might benefit from an activity identifying types of resources. Write a random list of energy resources and material resources for the class. Ask volunteers to identify each resource as either an energy resource or a material resource. Be sure to emphasize Section 3, "Using Material Resources," in Chapter 4, "Material Resources." Students must be able to identify different natural energy and material resources in order to master standard 6.6.b.

Question 9 *asks students to identify the product that is made from chemicals separated out of petroleum.*

A Incorrect. Glass is made from sand.

B Correct. Plastic is made from chemicals separated out of petroleum.

C Incorrect. Bricks are made from clay.

D Incorrect. Paper is made from wood.

Diagnostic Teaching Tip: Students who have difficulty answering this question correctly should view items made from different resources. Gather several items made from resources, such as a piece of paper, a brick, a glass vase, a wool sweater, a plastic item, a rubber eraser, a metal bolt, and so on. Have students guess what was used to make each one. Be sure to emphasize Section 3, "Using Material Resources," in Chapter 4, "Material Resources." Students must know the natural origin of the materials used to make common objects in order to master standard 6.6.c.

Question 10 *asks students to identify a situation in which natural resources are wasted.*

A Incorrect. Turning off the water while brushing your teeth conserves water.

B Correct. Running the dishwasher after every meal does not conserve resources.

C Incorrect. Recycling newspaper, glass, and aluminum cans conserves resources.

D Incorrect. Making sure the washing machine is full before you start it conserves energy and water.

Diagnostic Teaching Tip: Students who have difficulty with this question should benefit from a discussion about conserving natural resources. Discuss the various ways that students use natural resources at home and at school. Have students make a list of things that they already do or might do at home and at school to conserve natural resources. Be sure to emphasize Section 1, "Natural Resources," in Chapter 4, "Material Resources." Students must be able to identify different natural energy and material resources, and realize the importance of conserving them, in order to master standard 6.6.b.

Chapter Pretest

______ 1. Where is most of Earth's water located?
   A in the oceans, as salt water
   B in groundwater springs, as mineral water
   C in rivers and lakes, as fresh water
   D in glaciers, as fresh water

______ 2. Which of the following is NOT a source of fresh water?
   A a private well
   B the San Francisco Bay
   C the San Joaquin River
   D Lake Tahoe

______ 3. Which of the following natural resources would be best used to make electrical wiring?
   A a plant, such as a cacao tree
   B a nonmetal, such as carbon
   C a mineral, such as clay
   D a metal, such as copper

______ 4. Which of the following is NOT a process used to make rock?
   A cementation
   B pressure
   C cooling
   D evaporation

______ 5. How is energy transferred through an ecosystem?
   A Heat energy is changed into chemical energy by animals. The energy is transferred to plants when animals eat plants.
   B Kinetic energy is changed into chemical energy by plants. The energy is transferred to animals when they eat plants.
   C Sunlight is changed into chemical energy by plants. The energy is transferred to animals when they eat plants.
   D Mechanical energy is changed into chemical energy by animals. The energy is transferred to plants when animals eat plants.

______ 6. Which of the following is a naturally formed solid that has a consistent chemical composition and an orderly internal structure?
   A magma
   B ore
   C mineral
   D rock

Chapter Pretest *continued*

_______ **7.** Which of the following is a common mineral used to make computer chips?
  A copper
  B diamond
  C halite
  D quartz

_______ **8.** Which of the following is NOT a source of material resources?
  A the sun
  B the atmosphere
  C the ocean
  D organisms

_______ **9.** Which of the following is made from chemicals separated out of petroleum?
  A glass
  B plastic
  C bricks
  D paper

_______ **10.** Which of the following is NOT an example of conserving natural resources?
  A turning off the lights when you are not using them
  B running the dishwasher after every meal
  C recycling newspaper, glass, and aluminum cans
  D making sure the washing machine is full before you start it

Section Quiz

Section: Natural Resources

Match the correct definition with the correct term. Write the letter in the space provided.

_______ **1.** a natural resource that can be replaced at the same rate at which the resource is consumed

_______ **2.** any natural material that is used by humans, such as water, petroleum, minerals, forests, and animals

_______ **3.** a resource that forms at a rate that is much slower than the rate at which the resource is consumed

a. nonrenewable resource

b. renewable resource

c. natural resource

Write the letter of the correct answer in the space provided.

_______ **4.** Which of the following is NOT a renewable resource?
  a. coal
  b. water
  c. animal
  d. forest

_______ **5.** What is a process of reusing materials from waste or scrap called?
  a. reducing
  b. renewing
  c. recycling
  d. reclamation

_______ **6.** Examples of nonrenewable resources include
  a. wildlife.
  b. fresh water and air.
  c. trees.
  d. petroleum and natural gas.

_______ **7.** What is the purpose of recycling?
  a. to use natural resources
  b. to conserve energy
  c. to increase trash
  d. to use more energy

Assessment

Section Quiz

Section: Rock and Mineral Resources

Match the correct description with the correct term. Write the letter in the space provided.

_______ **1.** the process of returning land to its original condition after mining

_______ **2.** the removal of minerals that are located at or near Earth's surface

_______ **3.** the removal of ore that is deep within Earth

_______ **4.** the changes in pressure, temperature, or chemical makeup

a. surface mining

b. reclamation

c. subsurface mining

d. metamorphism

Write the letter of the correct answer in the space provided.

_______ **5.** A naturally formed inorganic solid with a crystalline structure is
  a. coal.
  b. volcanic glass.
  c. a chemical.
  d. a mineral.

_______ **6.** Types of surface mines include open pit, strip, and
  a. quarry.
  b. copper.
  c. subsurface.
  d. ore.

_______ **7.** Which of the following terms is used to describe a deposit that is mined for profit?
  a. mineral   **c.** ore
  b. stone   **d.** quarry

_______ **8.** Metals have shiny surfaces and
  a. are used to make cement.
  b. are good conductors of heat.
  c. are translucent.
  d. are good insulators of electricity.

_______ **9.** Which of the following materials is used to make aluminum?
  a. gypsum
  b. silica
  c. gold
  d. bauxite

Section Quiz

Section: Using Material Resources

Write the letter of the correct answer in the space provided.

______ **1.** From where can all material resources be obtained?
- **a.** a store
- **b.** Earth
- **c.** the sun
- **d.** computers

______ **2.** From where can nitrogen and oxygen be obtained?
- **a.** computers
- **b.** newspapers
- **c.** cotton
- **d.** the atmosphere

______ **3.** From where can petroleum be obtained?
- **a.** Earth's crust
- **b.** the oceans
- **c.** living things
- **d.** the atmosphere

______ **4.** From where can leather and dairy products be obtained?
- **a.** Earth's crust
- **b.** the oceans
- **c.** living things
- **d.** the atmosphere

______ **5.** What is the economic cost of producing an object?
- **a.** the price paid for an object at the store
- **b.** the total cost of making a product
- **c.** the cost of restoring damaged habitats
- **d.** the cost of the material resources used in a product

______ **6.** What is the environmental cost of producing an object?
- **a.** damage to an object after purchase
- **b.** the cost of making a product
- **c.** damage to land and habitats
- **d.** the cost of transporting a product

______ **7.** Natural resources that humans use to make objects or that are consumed are called
- **a.** material resources.
- **b.** atmospheric resources.
- **c.** energy resources.
- **d.** organic resources.

Chapter Test A

Material Resources

MULTIPLE CHOICE

Write the letter of the correct answer in the space provided.

______ **1.** Which of the following describes a renewable resource?
- **a.** A renewable resource is made from chemicals.
- **b.** A renewable resource takes thousands of years to form.
- **c.** A renewable resource can be replaced at the same rate it is used.
- **d.** A renewable resource can never be replaced.

______ **2.** Which of the following is a nonrenewable resource?
- **a.** forest
- **b.** wind
- **c.** fresh water
- **d.** natural gas

______ **3.** What are air and soil?
- **a.** polymers
- **b.** natural gases
- **c.** natural resources
- **d.** nonrenewable resources

______ **4.** Which of the following is NOT a good way to conserve energy?
- **a.** Turn off the lights.
- **b.** Run the washing machine half full.
- **c.** Take the bus.
- **d.** Ride a bike.

______ **5.** Which of the following describes a nonrenewable resource?
- **a.** A nonrenewable resource forms slowly.
- **b.** A nonrenewable resource can be replaced quickly.
- **c.** A nonrenewable resource is eaten by humans.
- **d.** A nonrenewable resource is found in Earth's atmosphere.

______ **6.** Which of the following are mined from the ground so they can be made into objects?
- **a.** rocks and minerals
- **b.** quarries
- **c.** organisms
- **d.** wood chips

Chapter Test A *continued*

_______ **7.** Which of the following do humans use to make objects?
    a. atmospheric resources
    b. artificial resources
    c. energy resources
    d. material resources

_______ **8.** If trees are cut down to make paper, how can we lower the environmental cost?
    a. by replanting the trees
    b. by cutting down more trees
    c. by improving manufacturing
    d. by making more paper

_______ **9.** Which of the following is the economic cost of a product?
    a. the process of reclamation
    b. the total cost of making the product
    c. the damage to the land
    d. the cost of pollutants

_______ **10.** Why do we recycle?
    a. to have more petroleum
    b. to increase trash
    c. to conserve energy
    d. to use natural resources

_______ **11.** Which of the following is NOT a renewable resource?
    a. coal
    b. tree
    c. air
    d. wind

_______ **12.** What can animal wastes be used for?
    a. to fertilize crops
    b. to make clothes
    c. to provide leather
    d. to make glass

_______ **13.** Which of the following is NOT provided by plants?
    a. fruit
    b. cotton
    c. lumber
    d. eggs

_______ **14.** What do we harvest from sea water?
    a. petroleum
    b. polymers
    c. salt
    d. aluminum

_______ **15.** Which of the following happens when there is a pressure and temperature change in rocks?
   a. reclamation
   b. metamorphism
   c. deposition
   d. evaporation

MATCHING

Match the correct description with the correct term. Write the letter in the space provided.

_______ **16.** a liquid mixture of hydrocarbons

_______ **17.** a polymer

_______ **18.** the most important resource in the atmosphere

   a. oxygen
   b. petroleum
   c. plastic

Match the correct description with the correct term. Write the letter in the space provided.

_______ **19.** reusing waste or scrap materials

_______ **20.** removing ore from Earth's surface

_______ **21.** digging tunnels deep within Earth

   a. surface mining
   b. subsurface mining
   c. recycling

FILL-IN-THE-BLANK

Use the terms from the following list to complete the sentences below.

   ore   reclamation
   mineral   rock

22. A naturally formed solid with a crystalline structure is called

   a(n) _______________________.

23. A deposit that is mined for profit is called a(n) _______________________.

24. The solid part of Earth's surface is made up of _______________________.

25. We can return land used for mining to its original state

   through _______________________.

Chapter Test B

Material Resources

MULTIPLE CHOICE

Write the letter of the correct answer in the space provided.

_______ **1.** What is any natural material used by humans called?
   a. a biological resource
   b. a nonrenewable resource
   c. a renewable resource
   d. a natural resource

_______ **2.** Which of the following is a renewable resource?
   a. fresh water
   b. coal
   c. petroleum
   d. natural gas

_______ **3.** Humans can conserve natural resources by
   a. leaving the lights on.
   b. relocating endangered species.
   c. keeping water sources clean.
   d. leaving the faucet on.

_______ **4.** Petroleum and natural gas are considered nonrenewable resources because
   a. they take thousands of years to form.
   b. they can be replaced quickly.
   c. they are made from chemicals.
   d. they help reduce pollution.

_______ **5.** What happens during deposition?
   a. A body of salt water dries up.
   b. Surface water carries dissolved materials into lakes.
   c. Groundwater is heated by magma.
   d. Magma cools slowly, and mineral crystals form.

_______ **6.** Metals that can be pounded into shapes are useful in making
   a. glass.
   b. fertilizer.
   c. bicycles.
   d. cement.

Chapter Test B *continued*

_______ **7.** Material resources are needed by humans
- **a.** to increase trash.
- **b.** to pollute Earth.
- **c.** to breathe.
- **d.** to make objects.

_______ **8.** Which of the following is the most valuable resource in the atmosphere?
- **a.** oxygen
- **b.** argon
- **c.** nitrogen
- **d.** carbon dioxide

_______ **9.** If the environmental cost of paper production is taken into account,
- **a.** old growth forests may disappear.
- **b.** local habitats may be disrupted.
- **c.** the economic cost may increase.
- **d.** the economic cost may be reduced.

_______ **10.** Earth's crust, oceans, atmosphere, and organisms are sources of
- **a.** unnatural resources.
- **b.** natural resources.
- **c.** economic resources.
- **d.** artificial resources.

_______ **11.** In subsurface mining,
- **a.** tunnels are dug into the ground.
- **b.** coal is strip mined.
- **c.** explosives are used.
- **d.** crushed rock is mined from quarries.

MATCHING

Match the correct description with the correct term. Write the letter in the space provided.

_______ **12.** makes up most of Earth's solid surface   **a.** ore

_______ **13.** is mined for profit   **b.** petroleum

     c. rock

_______ **14.** is used to produce fuel

| Chapter Test B *continued*

Match the correct description with the correct term. Write the letter in the space provided.

_______ **15.** reducing the amount of natural resources that must be obtained from Earth

_______ **16.** returning mined land to its original state

_______ **17.** removing coal from surface mines

a. strip mining

b. reclamation

c. recycling

Write the letter of the correct answer in the space provided.

_______ **18.** In addition to reclamation, a good way to reduce the environmental effects of mining is to
  a. recycle mineral products.
  b. dig deeper mines.
  c. use more metals.
  d. flush mines with water.

_______ **19.** Which of the following is NOT true of minerals?
  a. They are liquids.
  b. They are formed in nature.
  c. They are inorganic.
  d. They have a crystalline structure.

_______ **20.** What kinds of mines are open pit and quarry mines?
  a. shaft mines
  b. slope mines
  c. surface mines
  d. subsurface mines

Chapter Test B *continued*

MATCHING

Match the labels to the drawing. Write the letters in the spaces provided.

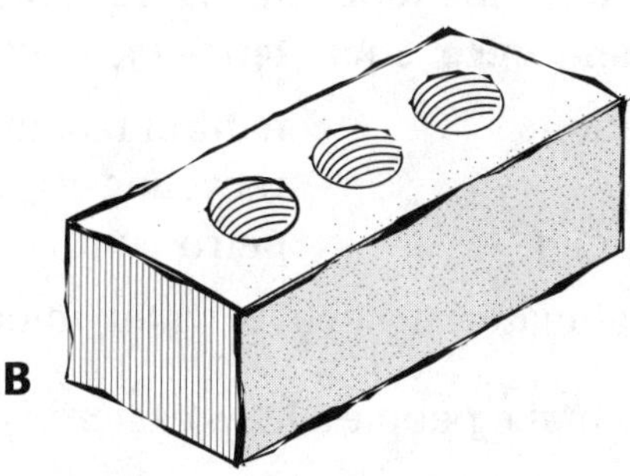

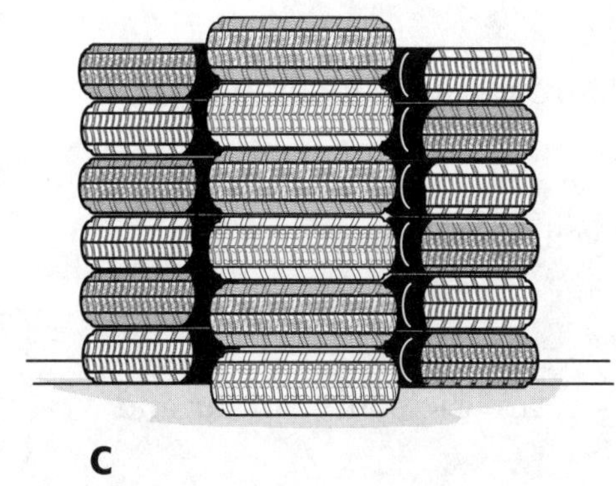

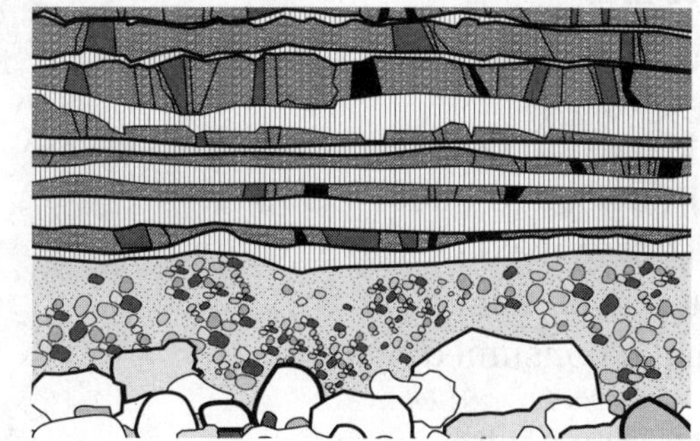

_______**21.** sand

_______**22.** clay

_______**23.** limestone

_______**24.** wool

_______**25.** rubber

Chapter Test C

Material Resources

USING KEY TERMS

Use the terms from the following list to complete the sentences below. Each term may be used only once. Some terms may not be used.

reclamation	natural resource	ore
rock	recycling	renewable
strip mining	petroleum	nonrenewable

1. Any natural material, such as water, minerals, and wildlife, that is used by

humans to make people's lives easier is called a(n) _____________________.

2. Natural resources that can be replaced in a short period of time are

_____________________ resources.

3. The process of recovering valuable materials from waste or scrap is

called _____________________.

4. A natural resource that can take millions of years to form is a(n)

_____________________ resource.

5. Some compounds in _____________________ are the source of tar and

asphalt.

UNDERSTANDING KEY IDEAS

Write the letter of the correct answer in the space provided.

_______ **6.** The term ore is used to describe a(n)
  a. plant resource used to make fertilizer.
  b. material resource used to make clothing.
  c. deposit that can be mined for profit.
  d. material that is consumed by humans.

_______ **7.** Humans make objects with
  a. material resources.
  b. human resources.
  c. energy resources.
  d. atmospheric resources.

_______ **8.** Strip mining is used
  a. when coal is found deep in Earth.
  b. in order to protect the environment.
  c. to limit the use of explosives.
  d. to mine shallow coal deposits.

Chapter Test C *continued*

______ **9.** Which of the following is NOT a mineral?
  a. gold
  b. calcite
  c. gypsum
  d. coal

______ **10.** The potentially harmful effects of mining can be reduced by
  a. reclamation of the land.
  b. strip mining.
  c. using more metals.
  d. flushing mines with water.

______ **11.** Which of the following is NOT made from gypsum?
  a. fertilizer
  b. cement
  c. plaster board
  d. plaster of Paris

12. List three ways in which humans can conserve natural resources, and give an example of each.

____

____

____

____

____

13. Name two methods of mining minerals, and explain when each method might be used.

____

____

____

____

____

____

____

14. List the five ways that minerals form, and explain what happens during each process.

____

____

____

____

____

CRITICAL THINKING

15. How can mining harm the environment?

____

____

____

16. List four economic costs that would affect how much a manufacturer would charge for paper, along with two potential environmental costs that could also affect the cost of the paper.

____

____

____

____

____

17. Name two examples of renewable natural resources that can still be used up too quickly, and explain how shortages of these resources occur.

____

____

____

____

| Chapter Test C *continued*

CONCEPT MAPPING

18. Use the following terms to complete the concept map below:

clothing	organisms	crust	atmosphere
minerals	oceans	salt	oxygen

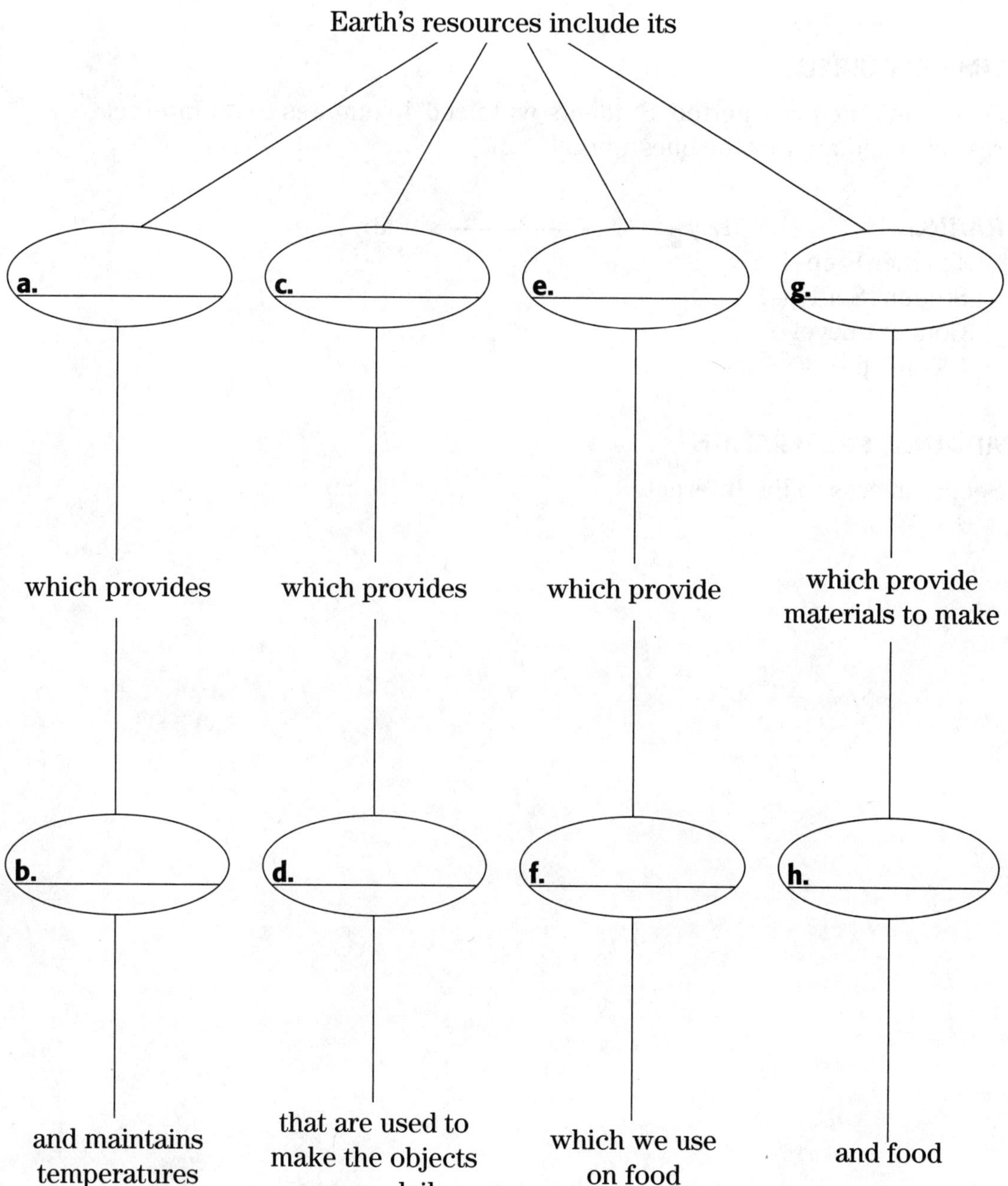

Assessment

Performance-Based Assessment

Teacher Notes

PURPOSE

Mining gives us the resources we need, but it can have an impact on our environment. Students will learn about the processes involved in mining a natural resource and the effect it has on our environment.

TIME REQUIRED

One 45-minute class period. Students will need 45 minutes to do Internet research and answer the questions.

RATING

Easy ← 1 2 3 4 → Hard

Teacher Prep–1
Student Set-Up–1
Concept Level–2
Clean Up–1

ADVANCE PREPARATION

Secure access to the Internet.

Performance-Based Assessment *continued*

Evaluation Strategies

Use the following rubric to help evaluate student performance.

Rubric for Assessment

Possible points	Analysis (100 points possible)
100–80	Clear, detailed answers; analysis stated clearly and accurately
79–50	Complete answers, but expressed in moderate or unclear manner
49–1	Erroneous or incomplete answers

Name _______________________ Class _______________ Date _______________

Performance-Based Assessment

OBJECTIVE

Learn how gold is found in the ground and mining's environmental impact.

KNOW THE SCORE!

As you work through the activity, keep in mind that you will be earning a grade for the following:

• how thoroughly you complete the activity (100%)

MATERIALS AND EQUIPMENT

• Internet access

PROCEDURE

1. Imagine that you are a prospector looking for gold. How do you find the gold?

____

____

____

2. How is gold extracted from a mine?

____

____

____

3. What is done with the gold after it is mined?

____

____

____

4. What is the impact of mining on the topsoil and wildlife?

____

____

____

____

Performance-Based Assessment *continued*

5. How can we reduce the harmful effects of gold mining?

____

____

____

____

6. How do humans use gold?

____

____

____

____

BIG IDEA QUESTION

7. Does the usefulness of gold outweigh the disturbance to the landscape? Explain your answer.

____

____

____

____

____

Assessment)

Standards Assessment

Teacher Notes and Answer Key

To provide practice under more realistic testing conditions, give students 20 min to answer all of the questions in this assessment.

QUESTION NUMBER	CORRECT ANSWER	STANDARD
1	*D*	6.7.b (supporting)
2	*B*	6.7.b (supporting)
3	*C*	6.7 (supporting)
4	*A*	6.7 (supporting)
5	*B*	6.6 (supporting)
6	*C*	6.6.b (mastering)
7	*D*	6.6.b (exceeding)
8	*C*	6.6.b (exceeding)
9	*B*	6.6.c (mastering)
10	*B*	6.6.b (exceeding)
11	*A*	6.6.b (supporting)
12	*A*	6.6.b (supporting)
13	*C*	6.6.c (exceeding)
14	*C*	5.2.f (mastering)
15	*D*	5.3.d (mastering)
16	*A*	4.4.b (mastering)

TEST DOCTOR

The following Standards Assessment questions have been diagnosed by the Test Doctor. Find out what might be causing your students' "ailing" answers. Each Test Doctor is followed by a diagnostic teaching tip to help you address students' learning needs.

Question 1 *asks students to identify the correct form of the word* select.

A Incorrect. The noun *selection* does not fit in this sentence.

B Incorrect. The adjective *selective* is not correct.

C Incorrect. The past tense, *selected,* cannot accompany *must* without the helping verb *have.*

D Correct. Scientists must *select* tools for their investigations.

> **Diagnostic Teaching Tip:** Students who have difficulty with this question might benefit from using each of the answer selections in a sentence. For example, have students revise the sentence in this question in ways that make each answer choice correct.

Question 2 *asks students to identify the definition of the word* technology.

A Incorrect. Science is the gathering of knowledge not the application of knowledge or tools.

B Correct. Technology is the application of knowledge or tools.

C Incorrect. Information is a kind of knowledge, not its application.

D Incorrect. Experimentation might lead to knowledge but is not the application of it.

Diagnostic Teaching Tip: Students who have difficulty answering this question correctly should practice defining words. Have students write short paragraphs describing how knowledge is gathered and applied using each of the answer choices given. Then, have students compare their descriptions with the dictionary definitions of these words.

Question 3 *asks students to choose the correct definition of* idea.

A Incorrect. An idea is not necessarily a hypothesis.
B Incorrect. Evidence is not an idea.
C Correct. An idea is a concept.
D Incorrect. An idea is not necessarily a prediction.

Diagnostic Teaching Tip: Students who struggle with this question might not be considering the nuances of each word. Write the word *idea* in a circle. Remind students that a hypothesis is a type of idea, and create a web containing *hypothesis, evidence,* and *prediction.* Have students brainstorm other kinds of ideas and add them to the web. Then, draw a horizontal line from *idea* to another circle containing the word *concept* and explain that the word is a synonym for *idea* rather than an example of a type of idea.

Question 4 *asks students to define the word* conduct *in a given context.*

A Correct. In this sentence, conduct means "to carry out or do."
B Incorrect. *Conduct* does not mean "to transmit or send" in this sentence.
C Incorrect. *Conduct* is not a manner of action in this sentence.
D Incorrect. In this sentence, *conduct* is a verb, not a noun meaning "standard of behavior."

Diagnostic Teaching Tip: Students who struggle with this question might benefit from taking the word *conduct* out of the given sentence and replacing it with the answer choices. This exercise should immediately eliminate answers C and D and lead to answer A.

Question 5 *asks students to choose a synonym for the word* required.

A Incorrect. *Required* does not have the same meaning as desired.
B Correct. Something that is needed can also be said to be required.
C Incorrect. The word *intended* does not mean the same thing as *required.*
D Incorrect. *Required* does not mean *available.*

Diagnostic Teaching Tip: Students who have trouble with this question might benefit from using the answer choices in context. For example, have students write a paragraph about their own lives that uses the words *desire, intend, available, necessary,* and *require.*

| Standards Assessment *continued*

Question 6 *asks students to demonstrate an understanding of the differences between renewable and nonrenewable resources.*

A Incorrect. Wind is a renewable resource.

B Incorrect. Though trees can be overused, they are considered renewable.

C Correct. Natural gas is a nonrenewable resource because it cannot replenish itself in a lifetime.

D Incorrect. Sunlight is a renewable resource.

Diagnostic Teaching Tip: Students who have difficulty with this question might benefit from creating a chart that shows how long it takes to create different resources. Note that resources that take much longer than a lifetime to renew are considered nonrenewable.

Question 7 *asks students to demonstrate an understanding of the criteria that identify a mineral.*

A Incorrect Minerals occur naturally.

B Incorrect. Minerals have a crystalline structure.

C Incorrect. A mineral is made of inorganic material.

D Correct. The chemical composition of a mineral is consistent throughout the material.

Diagnostic Teaching Tip: Students who have difficulty with this question might benefit from reviewing the four questions that identify a mineral while classifying samples of organic and inorganic material. Remind students that a negative answer to any question means that a material is not a mineral.

Question 8 *asks students to demonstrate an understanding of the differences between renewable and nonrenewable resources.*

A Incorrect. Renewable resources are not converted into nonrenewable resources.

B Incorrect. Renewable resources can be just as useful as nonrenewable resources.

C Correct. Many renewable resources, such as trees, can become scarce if used too quickly.

D Incorrect. Renewable resources can remain indefinitely if they are used correctly.

Diagnostic Teaching Tip: Students who have difficulty with this question might benefit from a review of which resources qualify as renewable. Have students write a paragraph explaining the benefits of several specific renewable resources.

Standards Assessment *continued*

Question 9 *asks students to demonstrate an understanding of the differences between organic and inorganic material.*

A Incorrect. Rocket fuels are made from oxygen in the atmosphere, not living things.

B Correct. Rope is made from plant fibers.

C Incorrect. Copper wiring is made from metal, which is inorganic.

D Incorrect. Plastics are made from petroleum, an inorganic material from Earth's crust.

Diagnostic Teaching Tip: Students who have difficulty with this question might benefit from noting the organic or inorganic origins of everyday objects. Have students brainstorm a list of items they use frequently and research what resources they are made from.

Question 10 *asks students to demonstrate knowledge of how gathering nonrenewable resources can harm the environment.*

A Incorrect. Reclamation does not reduce the amount of minerals used.

B Correct. Reclamation is a process of returning mines to their natural state.

C Incorrect. Minerals are not mined more efficiently in reclamation.

D Incorrect. Reclamation does not refer to where mining occurs.

Diagnostic Teaching Tip: Students who have difficulty with this question might benefit from reviewing the definitions of reclamation in the dictionary and discussing which definitions relate to the reclamation of mining areas.

Question 11 *asks students to identify common ways minerals are formed.*

A Correct. Minerals can be left behind when salt water evaporates.

B Incorrect. Minerals are inorganic, and materials formed from living organisms are generally organic.

C Incorrect. Manufacturers use minerals, but minerals are generally obtained from natural sources.

D Incorrect. This is not a common way that minerals form.

Diagnostic Teaching Tip: Students who have difficulty with this question might benefit from working with a group to describe a common way minerals form. Assign each group member a process by which minerals are formed, and have them explain that method to the group.

Question 12 *asks students to demonstrate an understanding of the formation of minerals.*

A Correct. The metallic elements form a crystalline structure, creating a new mineral.

B Incorrect. The elements do not dissolve to form a mineral.

C Incorrect. Minerals are not made of organic materials.

D Incorrect. Minerals must have a consistent chemical composition.

Standards Assessment *continued*

Diagnostic Teaching Tip: Students who have difficulty with this question might benefit from a review of the criteria for defining minerals. Have students create a checklist for three different minerals. Instruct them to check off criteria as you review the way in which each mineral is formed.

Question 13 *asks students to recognize a hidden environmental cost associated with plastic products.*

A Incorrect. Manufacturers include the cost of raw materials in the price of a product.

B Incorrect. Manufacturing costs are taken into account when setting the price for a given product.

C Correct. Costs that are not paid directly by the manufacturing company, such as damage to the environment, are not included in the price of a product.

D Incorrect. Advertising is a cost that manufacturers include in the price of their products.

Diagnostic Teaching Tip: Students who have difficulty with this question should be reminded that manufacturers, like all for-profit businesses, are trying to make a profit by selling their product for more than it costs them to make the product. Certain costs, like damage to the environment, are considered "hidden" because they are not paid by the manufacturer directly. Have students consider the hidden costs associated with various businesses, such as manufacturing, mining, and farming.

Question 14 *asks students to demonstrate knowledge of the process of photosynthesis.*

A Incorrect. Carbon dioxide is released by plants during cellular respiration; however, carbon dioxide is not a product of photosynthesis.

B Incorrect. Carbon monoxide is not a resource that plants release.

C Correct. Plants release oxygen into the atmosphere. This by-product of photosynthesis is an important resource for other organisms.

D Incorrect. Nitrogen is not a resource plants release into the atmosphere.

Diagnostic Teaching Tip: Students who have difficulty with this question might benefit from reviewing the process of photosynthesis. Have students create a flowchart illustrating the process of photosynthesis and how it benefits humans and other organisms on Earth.

Question 15 *asks students to demonstrate an understanding of the origins of fresh water in California.*

A Incorrect. Fresh water is not taken primarily from the ocean.

B Incorrect. Fresh water is a limited resource and should be conserved.

C Incorrect. Fresh water will not disappear if it is conserved.

D Correct. Fresh water is a renewable resource, but it is limited and must be conserved.

Standards Assessment *continued*

Diagnostic Teaching Tip: Students who have difficulty with this question might benefit from reviewing fresh water sources in their area. Help students investigate how much fresh water is available and how this source is replenished.

Question 16 *asks students to demonstrate an understanding that minerals can be classified by their properties.*

A Correct. A mineral's hardness is measured by the Mohs scale.
B Incorrect. The Mohs scale does not measure color.
C Incorrect. Electrical conductivity is not measured by the Mohs scale.
D Incorrect. The Mohs scale does not measure structure.

Diagnostic Teaching Tip: Students who have difficulty with this question might benefit from reviewing a chart showing the values of several minerals on the Mohs hardness scale. After reviewing the chart, pair two minerals, and invite students to determine which mineral would be scratched if the two were scraped against each other.

Assessment

Standards Assessment

REVIEWING ACADEMIC VOCABULARY

______ **1.** Which of the following words best completes the following sentence:
"Scientists must _____ the correct tools for their experiments"?
A selection
B selective
C selected
D select

______ **2.** Which of the following words means "the application of science for
practical purposes"?
A science
B technology
C information
D experimentation

______ **3.** Which of the following words is closest in meaning to the word *idea*?
A hypothesis
B evidence
C concept
D prediction

______ **4.** In the sentence "Scientists conduct experiments to test new hypoth-
eses," what does the word *conduct* mean?
A to carry out or do
B to transmit or send
C manner of action
D standard of behavior

______ **5.** Which of the following words is closest in meaning to the word
required?
A desired
B needed
C intended
D available

REVIEWING CONCEPTS

______ **6.** Which of the following resources is nonrenewable?
A wind
B trees
C natural gas
D sunlight

Standards Assessment *continued*

_______ **7.** Which of the following observations is evidence that a material is not
a mineral?
  A The material occurs naturally.
  B The material has a crystalline structure.
  C The material is inorganic.
  D The material has a varied chemical composition.

_______ **8.** Which of the following is true of renewable resources?
  A They must be converted into nonrenewable resources.
  B They are less useful than nonrenewable resources.
  C Many of them can become scarce if used too quickly.
  D No matter how much we conserve, they will one day be gone.

_______ **9.** Which of the following products is made from resources provided by
plants?
  A rocket fuel
  B cotton rope
  C copper wiring
  D plastic

_______ **10.** How does reclamation reduce the harmful effects of mining?
  A by reducing the use of minerals
  B by returning mines to their natural state
  C by mining minerals more efficiently
  D by avoiding mining in protected areas

_______ **11.** Which of the following is a common way that minerals
are formed?
  A They are left behind when salt water evaporates.
  B They are formed from plant material that collects in swamps.
  C They are created by mineral manufacturers.
  D They are formed when the sun's rays interact with soil.

Standards Assessment *continued*

1 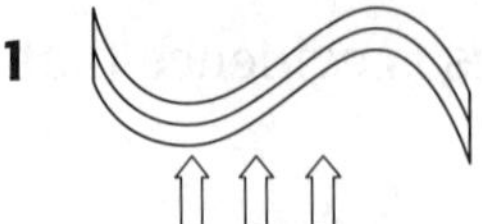
**Water is heated
by magma.**

2 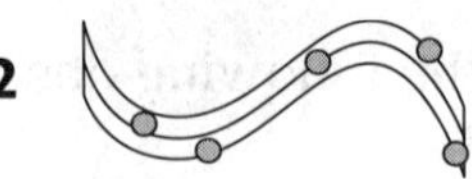
**Hot water and
metallic elements
form a solution.**

3 

4 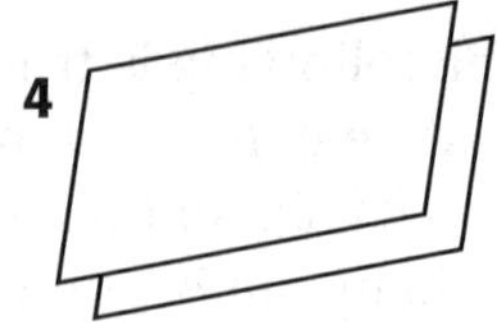
**New minerals
are formed.**

______ **12.** The diagram above shows one way in which new minerals are formed.
Which step is most likely missing from this process?
A The elements crystallize.
B The elements dissolve.
C The elements mix with organic compounds.
D The elements combine with other elements.

______ **13.** Which of the following costs might not be included in the price of a
plastic product?
A the cost of the raw material—petroleum
B the cost of manufacturing the product
C the environmental cost of drilling for petroleum
D the cost of advertising the product

REVIEWING PRIOR LEARNING

______ **14.** What is the product of photosynthesis represented by the arrows in
the diagram above?
A carbon dioxide
B carbon monoxide
C oxygen
D nitrogen

Standards Assessment *continued*

_______ **15.** Which of the following statements is true of fresh water used in California?

    A It is derived primarily from desalinated ocean water.

    B It is a renewable resource that can be used freely.

    C It will one day be gone in spite of conservation.

    D It is a limited resource that should be conserved.

_______ **16.** Which of the following is a physical property of minerals measured by the Mohs scale?

    A hardness

    B color

    C ability to conduct electricity

    D crystalline structure

What Is Your Classroom Made Of?

Teacher Notes

This activity helps students understand the natural origins of products they use daily (covers standard 6.6.c). Many students may be unaware that most plastics (also called *synthetic polymers*) are made from derivatives of petroleum, such as the hydrocarbons benzene and ethylene. Other resources, such as glass fibers or carbon fibers, may be added to reinforce plastics that need to be very strong. If students are unable to name the natural origin for plastic, provide this information.

MATERIALS

For each group

• paper

Explore Activity ）   **DATASHEET A**

What Is Your Classroom Made Of?

Complete the following activity to see if you can figure out the natural origin, or source, of materials used to make common objects in your classroom.

PROCEDURE

1. On a **sheet of paper,** make three columns.
   - Label the first column "Object."
   - Label the second column "Parts of the object."
   - Label the third column "Natural origin."

2. Look around your classroom.
   - Choose a few objects to put on your list under "Object."

3. With a partner, discuss each item that you have chosen.
   - List the parts of each object in the second column on your paper.
   - For example, for the object "window," parts of the window could be "glass" and "frame."

ANALYSIS

4. With your partner, discuss the natural origin of the parts of each object.
   - Add the natural origins to the third column of your table.
   - For example, for "glass" the natural origin could be "sand," and for "frame" the natural origin could be "trees" or "metal ore."

5. If you can, describe how one of the objects on your list is made.

____

____

____

6. For one object on your list, list all the economic costs you can think of that are included in that object's purchase price. (Hint: The costs that go into making that object are only part of the purchase price. What are some costs that might go into it after it is made?)

____

____

____

____

What Is Your Classroom Made Of?

Complete the following activity to see if you can determine the natural origin of materials used to make common objects in your classroom.

PROCEDURE

1. On a **sheet of paper,** make three columns. Label one column "Object," the second column "Parts of the object," and the third column "Natural origin."

2. Look around your classroom. Choose a few objects to put on your list under "Object."

3. With a partner, discuss each item that you have chosen. List the parts of each object in the second column on your paper. For example, for the object "window," parts of the window could be "glass" and "frame."

ANALYSIS

4. With your partner, discuss the natural origin of the parts of each object and add the natural origins to the third column of your table. For example, for "glass," the natural origin could be "sand."

5. If you can, describe the manufacturing process for some of the objects on your list.

____

____

6. What do you think are the costs that are included in the purchase price of some of the objects on your list?

____

____

____

____

____

____

Explore Activity   **DATASHEET C**

What Is Your Classroom Made Of?

Complete the following activity to see if you can determine the natural origin of materials used to make common objects in your classroom.

PROCEDURE

1. On a **sheet of paper,** make three columns. Label one column "Object," the second column "Parts of the object," and the third column "Natural origin."

2. Look around your classroom. Choose a few objects to put on your list under "Object."

3. With a partner, discuss each item that you have chosen. List the parts of each object in the second column on your paper. For example, for the object "window," parts of the window might be "glass" and "frame."

ANALYSIS

4. With your partner, discuss the natural origin of the parts of each object, and add the natural origins to the third column of your table. For example, for "glass" the natural origin could be "sand."

5. Describe the manufacturing process for some of the objects on your list.

____

____

____

____

6. What economic costs are included in the purchase price of some of the objects on your list?

____

____

____

• What environmental costs might be included in the purchase price of some of the objects on your list?

____

____

DATASHEET

Renewable or Not?

Teacher Notes

This activity helps students understand how the rate of resource formation or replacement and the rate of resource use combine to determine whether a resource is renewable or nonrenewable (covers standard 6.6.b). After completing the activity, have students discuss the concrete, everyday actions they can take to reduce the "holes in the cup."

MATERIALS

For each group

- cups, paper (2)
- pencil, sharpened
- sink or basin
- water

SAFETY CAUTION

Remind students to review all safety cautions and icons before beginning this activity.

Renewable or Not?

SAFETY INFORMATION

PROCEDURE

1. Gather two **paper cups.**
   - Fill one cup 3/4 full with **water.**

2. Put three holes in the bottom of the second cup with a **sharpened pencil.**
   - Make a pencil mark on the inside of the second cup 2 cm from the bottom.

3. Hold the second cup over a **basin** or **sink.**
   - Pour water into the second cup.
   - Try to keep the water level even with the mark.

4. The faster you pour water into the cup, the _______________________ the water level in the cup is.

5. Which part of this model represents the use of a resource? Which part represents the formation of a resource?
   - The water coming out of the holes in the cup represents

   ____

   - The water poured into the cup represents

   ____

6. What makes a resource renewable?

   ____

   - What makes a resource nonrenewable?

   ____

   ____

Renewable or Not?

SAFETY INFORMATION

PROCEDURE

1. Gather two **paper cups,** and fill one 3/4 full with **water.**

2. Put three holes in the bottom of the second cup with a **sharpened pencil.** Make a pencil mark on the inside of the second cup about 2 cm from the bottom.

3. Hold the second cup over a **basin** or **sink.** Pour water into the second cup, and try to keep the water level even with the mark.

4. How does the rate at which you pour water into the cup relate to the water level in the cup?

____

____

5. Which part of this model represents the use of a resource? Which part represents the formation of a resource?

____

____

____

6. What makes a resource renewable? What makes a resource nonrenewable?

____

____

____

____

____

Quick Lab

DATASHEET C

Renewable or Not?

SAFETY INFORMATION

PROCEDURE

1. Gather two **paper cups,** and fill one 3/4 full with **water.**

2. Put three holes in the bottom of the second cup with a **sharpened pencil.** Make a pencil mark on the inside of the second cup about 2 cm from the bottom.

3. Hold the second cup over a **basin** or **sink.** Pour water into the second cup, and try to keep the water level even with the mark.

4. Name a factor that controls the level of water in the second cup.

____

____

____

5. How does this model represent the formation and the use of a resource?

____

____

____

6. What determines whether a resource is renewable or nonrenewable?

____

____

____

DATASHEET

Chocolate Ore

Teacher Notes

This activity helps students learn how the concentration of an ore is determined (covers standard 6.6.c). If different brands of cookies are used, label them so that students can compare percentages in question 6.

MATERIALS

For each student

- cookie, chocolate chip
- paper clips or toothpicks
- laboratory balance (shared)

SAFETY CAUTION

Remind students to review all safety cautions and icons before beginning this activity.

Name _______________________________ Class _______________ Date _____________

Chocolate Ore

Materials from Earth are mined only if the ore deposits are large enough to make the mining profitable. This activity demonstrates how miners find the concentration of an ore.

SAFETY INFORMATION

TRY IT!

1. Get a **chocolate chip cookie.**
   - Collect some "mining tools" such as **paper clips** and **toothpicks.**

2. Find the mass of your whole cookie on a **laboratory balance.** (Hint: Make sure to set the balance to zero before you start.)
   - Record the mass of the cookie. _______

3. Use your mining tools to mine the chocolate out of your cookie.
   - Separate the cookie from the chocolate chips.

4. Collect all of the chocolate chips from your cookie.
   - Determine the total mass of chocolate by using the balance.
   - Record the mass of the chocolate. _______

THINK ABOUT IT!

5. Find the percentage of chocolate ore in your cookie as follows:
   - Divide the mass of the chocolate by the mass of the whole cookie.
   - Multiply the result by 100.
   - This will give you the percentage of chocolate ore in your cookie.
   - Record the percentage of chocolate ore. _______

6. Compare your percentage of chocolate ore with the percentages of your classmates. Are some cookies richer in chocolate ore than others are? (Hint: "Richer in" means "containing more.")

____

____

____

7. Copper ore is rock that contains at least 0.7% copper. Pretend that the same percentage of chocolate is needed to make a cookie chocolate ore. Did your cookie have enough chocolate to make it chocolate ore? Explain your answer.

____

____

Quick Lab

DATASHEET B

Chocolate Ore

Materials from Earth are mined only if there are ore deposits concentrated enough to make the mining profitable. This activity demonstrates how miners find the concentration of an ore.

SAFETY INFORMATION

TRY IT!

1. Gather a **chocolate chip cookie** and some "mining tools" such as **paper clips** and **toothpicks.**

2. Determine the mass of your whole cookie on a **laboratory balance.** Record the mass.

____

3. Use your mining tools to mine the chocolate out of your cookie by separating the cookie from the chocolate chips.

4. Collect all of the chocolate chips from your cookie, and determine the total mass of the chocolate by using the balance.

____

THINK ABOUT IT!

5. Calculate the percentage of chocolate ore in your cookie by dividing the mass of the chocolate by the mass of the whole cookie and multiplying the result by 100. Show your work below.

Chocolate Ore *continued*

6. Compare your percentage of chocolate ore with the results of your classmates. Are some cookies richer in chocolate ore than others are?

____

____

7. What determines whether a deposit of a natural material is an ore? Did your cookie contain enough chocolate to be considered chocolate ore? Explain your answer.

____

____

____

____

____

Quick Lab

Chocolate Ore

Materials from Earth are mined only if there are ore deposits concentrated enough to make the mining profitable. This activity demonstrates how miners find the concentration of an ore.

SAFETY INFORMATION

TRY IT!

1. Gather a **chocolate chip cookie** and some "mining tools" such as **paper clips** and **toothpicks.**

2. Determine the mass of your whole cookie on a **laboratory balance.** Record the mass.

____

3. Use your mining tools to mine the chocolate out of your cookie by separating the cookie from the chocolate chips.

4. Collect all of the chocolate chips from your cookie, and determine the total mass of chocolate by using the balance.

____

THINK ABOUT IT!

5. Calculate the percentage of chocolate ore in your cookie. Show your work. below

6. Compare your percentage of chocolate ore with the results of your classmates.

____

____

____

____

Chocolate Ore *continued*

7a. What determines whether a deposit of a natural material is an ore? Would your cookie be considered chocolate ore? Explain your answer.

____

____

____

____

7b. What type of mining did you have to perform to get to your chocolate ore? Explain your answer.

____

____

____

____

7c. What are the environmental costs of your mining operation? Is there any hope of reclamation?

____

____

____

____

Products from Plants

Teacher Notes

This activity helps students understand the natural origin of products (covers standard 6.6.c). You may wish to have students wear gloves. Line an out-of-the-way shelf or ledge with newspaper where students can store their objects. Once the objects have dried, students could sand and paint them. Have students spread newspapers over their work area before they sand and paint.

MATERIALS

for Each Pair of Students

- bowl, small
- flour (½ c)
- sandpaper
- sawdust (1 c)
- starch, liquid (1 TBSP)
- water (1 c)

SAFETY CAUTION

Remind students to review all safety cautions and icons before beginning this activity.

DATASHEET A

Products from Plants

SAFETY INFORMATION

PROCEDURE

1. Mix 1 cup **sawdust**, 1/2 cup **flour, 1 TBSP liquid starch,** and **1 cup water** in a **small bowl.**
   - The dough should be fairly stiff.
   - If the dough is too dry, add more water.
2. Make an object with your dough.
   - Set the object aside to dry for 2 or 3 days.
   - After it is dry, use sandpaper to smooth the object's surface.
3. What is the natural origin of the materials you used?

   sawdust ____

   flour ____

   liquid starch ____

   water ____

   Are these resources renewable or nonrenewable?

   ____

   ____

   ____

   ____

4. What other objects are made from these materials?

   sawdust____

   flour ____

   liquid starch ____

   water ____

  DATASHEET B

Products from Plants

SAFETY INFORMATION

PROCEDURE

1. Mix **1 cup sawdust, 1/2 cup flour, 1 TBSP liquid starch,** and **1 cup water** in a **small bowl.** The dough should be fairly stiff, but if it is too dry, add more water.

2. Make an object with your dough, and set it aside to dry for 2 or 3 days. After it is dry, use **sandpaper** to smooth the surface.

3. What is the natural origin of the materials you used to make your object? Are these resources renewable or nonrenewable?

____

____

____

4. What other objects can you name that are made from the materials used here?

____

____

____

____

____

____

____

Products from Plants

SAFETY INFORMATION

PROCEDURE

1. Mix **1 cup sawdust, 1/2 cup flour, 1 TBSP liquid starch,** and **1 cup water** in a **small bowl**. The dough should be fairly stiff, but if it is too dry, add more water.

2. Make an object with your dough, and set it aside to dry for 2 or 3 days. After it is dry, use sandpaper to smooth the surface.

3. What is the natural origin of the materials you used to make your object? Are these resources renewable or nonrenewable? Explain your answer.

____

____

____

____

4. What other objects are manufactured from the materials used here? Give examples for each one.

____

____

____

____

Skills Practice Lab)

Natural Resources Used at Lunch

Teacher Notes

This activity has students determine the resources that are used to make packaging materials for food (covers standards 6.6.b, 6.6.c, and 6.7.b). Set up groups of four. If groups must share a balance, you may want to suggest that one group measure one category at a time and calculate the figures in the first two columns of the table while the second group uses the balance. This sequence may help prevent having groups wait for long periods of time to use the balance.

Dan Judnick
Bernal
Intermediate School
San Jose, California

TIME REQUIRED

One 45-minute class period

LAB RATINGS

Easy ←——— 1 2 3 4 ———→ Hard

Teacher Prep–1
Student Set-Up–3
Concept Level–3
Clean Up–2

MATERIALS

The materials listed on the student page are enough for 4 students.

SAFETY CAUTION

Students should wear gloves and aprons and avoid handling metal lids or containers by edges.

PREPARATION NOTES

You may want to demonstrate the use of the balance the day before students are to conduct the lab.

 DATASHEET A

Natural Resources Used at Lunch

Many materials in addition to food are involved in the packaging and delivery of lunches. These materials are made from lots of different natural resources. In this activity, you will see what types of resources are used in the packaging and delivery of lunches at your school.

OBJECTIVES

Categorize common materials left over after lunch.

Compute the amounts in each category, and determine the percentage of the total represented by each category.

Recommend ways your school can conserve natural resources based on the data you collect and organize in this activity.

MATERIALS

- bags, plastic
- balance, triple beam or electronic
- calculator
- gloves, protective, plastic
- paper towels

SAFETY INFORMATION 

Using Scientific Methods

ASK A QUESTION

1. What percentage of the materials left over after lunches comes from each of the following categories: paper and wood products, plastic, metal, and glass?

FORM A HYPOTHESIS

2. Complete the following hypothesis so that it answers the question above.

When we separate the materials left over from lunch, we will find _______ % are

paper and wood products, _______ % are plastic products, _______ % are materials

made of metal, and _______ % are materials made of glass.

| Natural Resources Used at Lunch *continued*

TEST THE HYPOTHESIS

3. Collect all of your lunch waste.
   - Collect all of your lunch waste on the day of the lab activity or the day before the lab activity, depending on whether your class meets before or after lunch.
   - Put all of your lunch waste in a plastic bag, including wrappers, napkins, straws, and disposable trays.

4. Working in groups of three or four students, separate your lunch waste onto paper towels in the following categories:
   - paper and wood
   - plastic
   - metal
   - glass

Data for Leftover Lunch Materials				
Category	Total mass for lab group	Average mass per student	Percentage of total waste	Notes
Paper and wood				
Plastic				
Metal				
Glass				
Total			100	

5. Using the balance, determine the mass of the waste in each category for the entire group.
   - Record the masses in the table above.

6. Use the equation below to calculate for each category the average mass of solid waste per student.
   - Take the total mass for paper and wood, and then divide it by the number of students in your lab group. This will give you the average mass of paper and wood waste per student.
   - Repeat this step for each category of waste.
   - Record the results in the table.

$$\frac{total\ mass\ in\ category}{number\ of\ students} = \begin{array}{l} average\ mass\ in \\ category\ per\ student \end{array}$$

Natural Resources Used at Lunch *continued*

7. Use the equation below to calculate the percentage of the total waste that is represented by each category.
  - Take the total mass for paper and wood, divide it by the total mass for all categories, and then divide that number by 100. This will give you the percentage of the total waste.
  - Repeat this step for each category of waste.
  - Record the results in the table.

$$\left(\frac{\text{total mass in category}}{\begin{array}{c} \text{total mass for} \\ \text{all categories} \end{array}} \right) 100 \;=\; \begin{array}{l} \text{percentage of} \\ \text{total waste} \end{array}$$

ANALYZE THE RESULTS

8. Examining Data Compare your group's percentages with the percentages of other groups in the class. How are the data similar?

____

____

____

How are the data different?

____

____

____

Why are some data similar?

____

____

____

Why are some data different?

____

____

____

____

Natural Resources Used at Lunch *continued*

9. Classifying In the "Notes" column of the table:
- List the natural origin of the materials in each category. The natural origin is where the materials first came from; for example, wood comes from trees.
- List whether each category of materials is made from renewable or nonrenewable resources.

DRAW CONCLUSIONS

10. Interpreting Information What percentage of these lunch leftover materials came from renewable resources? (Hint: If more than one category is a renewable resource, then add together the percentages for each category to get the total percentage of renewable resources.)

____

What percentage came from nonrenewable resources?

____

11. Making Predictions Find the mass of lunch materials in each category that are left over at your school each day.
- Ask your teacher how many people eat lunch at your school everyday.
- Record this number: _____________________________
- Multiply the number above with the mass per student (column two in the data table) in each category. Record your answers below.

Paper and wood _____________________________

Plastic _____________________________

Metal _____________________________

Glass _____________________________

BIG IDEA QUESTION

12. Applying Conclusions You have been asked to suggest some steps the school can take to conserve material resources.
- Write at least two suggestions.
- Base your suggestions on the data you have collected during this activity and on what you know about the material resources you have been studying.

____

____

____

____

____

Natural Resources Used at Lunch

Many materials in addition to food are involved in the packaging and delivery of lunches. These materials are made from a variety of natural resources. In this activity, you will determine what types of resources are used in this way at your school.

OBJECTIVES

Categorize common materials left over after lunch.

Compute the amounts in each category, and determine the percentage of the total represented by each category.

Recommend ways your school can conserve natural resources based on the data you collect and organize in this activity.

MATERIALS

- bags, plastic
- balance, triple beam or electronic
- calculator
- gloves, protective, plastic
- paper towels

SAFETY INFORMATION 

Using Scientific Methods

ASK A QUESTION

1. What percentage of the materials left over after lunch come from each of the following categories: paper and wood products, plastic, metal, and glass?

FORM A HYPOTHESIS

2. Write a hypothesis that is a possible answer to the question above. Explain your reasoning.

____

____

____

Natural Resources Used at Lunch *continued*

TEST THE HYPOTHESIS

3. Collect all of your lunch waste on the day of the lab activity or the day before the lab activity, depending on whether your class meets before or after lunch. Put all of your lunch waste in a plastic bag, including wrappers, napkins, straws, and disposable trays.

Data for Leftover Lunch Materials				
Category	Total mass for lab group	Average mass per student	Percentage of total waste	Notes
Paper and wood				
Plastic				
Metal				
Glass				
Total			100	

4. Working in groups of three or four students, separate your lunch waste onto paper towels in the following categories: paper and wood, plastic, metal, and glass.

5. Determine the mass of the waste in each category for the entire group. Create a data table similar to the one above, and record the masses.

6. Use the equation below to calculate for each category the average mass of solid waste per student. Use the total mass of waste for each category in your calculation. Record the results in your table.

$$\frac{total\ mass\ in\ category}{number\ of\ students} = \begin{array}{l} average\ mass\ in \\ category\ per\ student \end{array}$$

7. Use the equation below to calculate the percentage of the total waste that is represented by each category. Record the results in your table.

$$\left(\frac{total\ mass\ in\ category}{\begin{array}{c} total\ mass\ for \\ all\ categories \end{array}} \right) 100 = \begin{array}{l} percentage\ of \\ total\ waste \end{array}$$

Natural Resources Used at Lunch *continued*

ANALYZE THE RESULTS

8. Examining Data Compare your group's percentages for each category with the results from other groups in the class. How and why are the data similar or different?

____

____

9. Classifying In the "Notes" column of your table, list the natural origin of the materials in each category. List whether each category of materials is made from renewable or nonrenewable resources.

DRAW CONCLUSIONS

10. Interpreting Information What percentage of these lunch leftover materials came from renewable resources? What percentage came from nonrenewable resources?

____

____

11. Making Predictions Find the mass of leftover materials at your school in each category. Your calculations should be done to find the mass generated each day. You may need to ask your teacher how many students, teachers, and staff members there are at your school.

____

____

BIG IDEA QUESTION

12. Applying Conclusions You have been asked to recommend to your school some steps the school can take to conserve material resources. Describe at least two things you would recommend. Base your recommendations on the data you have collected during this activity and on what you know about the material resources you have been studying.

____

____

____

____

 DATASHEET C

Natural Resources Used at Lunch

Many materials in addition to food are involved in the packaging and delivery of lunches. These materials are made from a variety of natural resources. In this activity, you will determine what types of resources are used in this way at your school.

OBJECTIVES

Categorize common materials left over after lunch.

Compute the amounts in each category, and determine the percentage of the total represented by each category.

Recommend ways your school can conserve natural resources based on the data you collect and organize in this activity.

MATERIALS

- bags, plastic
- balance, triple beam or electronic
- calculator
- gloves, protective, plastic
- paper towels

SAFETY INFORMATION 

Using Scientific Methods

ASK A QUESTION

1. What percentage of the materials left over after lunch comes from each of the following categories: paper and wood products, plastic, metal, and glass?

FORM A HYPOTHESIS

2. Write a hypothesis that is a possible answer to the question above. Explain your reasoning.

____

____

____

____

____

| Natural Resources Used at Lunch *continued*

TEST THE HYPOTHESIS

3. Collect all of your lunch wastes on the day of the lab activity or the day before the lab activity, depending on whether your class meets before or after lunch. Put all of your lunch waste in a plastic bag, including wrappers, napkins, straws, and disposable trays.

4. Working in groups of three or four students, separate your lunch waste onto paper towels in the following categories: paper and wood, plastic, metal, and glass.

Data for Leftover Lunch Materials

Category	Total mass for lab group	Average mass per student	Percentage of total waste	Notes
Paper and wood				
Plastic				
Metal				
Glass				
Total			100	

5. Determine the mass of the waste in each category for the entire group. Create a data table similar to the one above, and record the masses.

6. Use the equation below to calculate for each category the average mass of solid waste per student. Use the total mass of waste for each category in your calculation. Record the results in your table.

$$\frac{total\ mass\ in\ category}{number\ of\ students} = \frac{average\ mass\ in}{category\ per\ student}$$

7. Use the equation below to calculate the percentage of the total waste that is represented by each category. Record the results in your table.

$$\left(\frac{total\ mass\ in\ category}{total\ mass\ for\ all\ categories}\right)100 = \frac{percentage\ of}{total\ waste}$$

ANALYZE THE RESULTS

8. Examining Data Compare your group's percentages with percentages of other groups in class. How and why are the data similar or different?

____

____

____

____

9. Classifying In the "Notes" column of your table, list the natural origin of the materials in each category. List whether each category of materials is made from renewable or nonrenewable resources.

DRAW CONCLUSIONS

10. Interpreting Information Compare the percentage of lunch leftover materials from renewable resources with the percentage from nonrenewable resources.

____

____

____

____

11. Making Predictions Find the mass of leftover materials in each category that are generated at your school in one day. Then, determine the total mass of waste in each category generated in a school year.

____

____

____

____

____

Natural Resources Used at Lunch *continued*

BIG IDEA QUESTION

12. Applying Conclusions You have been asked to recommend to your school some steps the school can take to conserve material resources. Describe at least two things you would recommend. Base your recommendations on the data you have collected during this activity and on what you know about the material resources you have been studying. How could your recommendations be put into practice?

____

____

____

____

____

____

____

Reading a Geologic Map

Teacher Notes

This activity explains features and symbols on geologic maps. Students are then asked to read a geologic map (covers standard 6.7.f). If possible, display a geologic map that includes your community to show students how the surface features with which they are familiar are shown on a geologic map.

Reading a Geologic Map

INVESTIGATION AND EXPERIMENTATION

6.7.f Read a topographic map and a geologic map for evidence provided on the maps and construct and interpret a simple scale map.

TUTORIAL

All Earth scientists use geologic maps. Geologic maps show the types of rocks found at the surface in a given area. They also show the locations of geologic structures such as faults and folds. The following list will help you read geologic maps.

1. Geologic maps are made on a base map that shows things such as topography, roads, and rivers. These features are usually shown in light colors or as gray lines.

2. Rock of the same age and type is called a *rock unit*. On geologic maps, each rock unit is represented by a different color. Different units that are similar in age are often represented by different shades of the same color.

3. Each rock unit is identified on a geologic map by a set of letters:
   - Capital letters indicate the age of the rock (for example, Q = Quaternary and T = Tertiary).
   - Lowercase letters indicate the type of rock (for example, al = alluvium, v = volcanic, and i = intrusive).

4. Other geologic features are indicated on the map by special symbols:
   - Faults are shown as thick black lines.
   - Contacts between rock units are shown as thin black lines.

Reading a Geologic Map *continued*

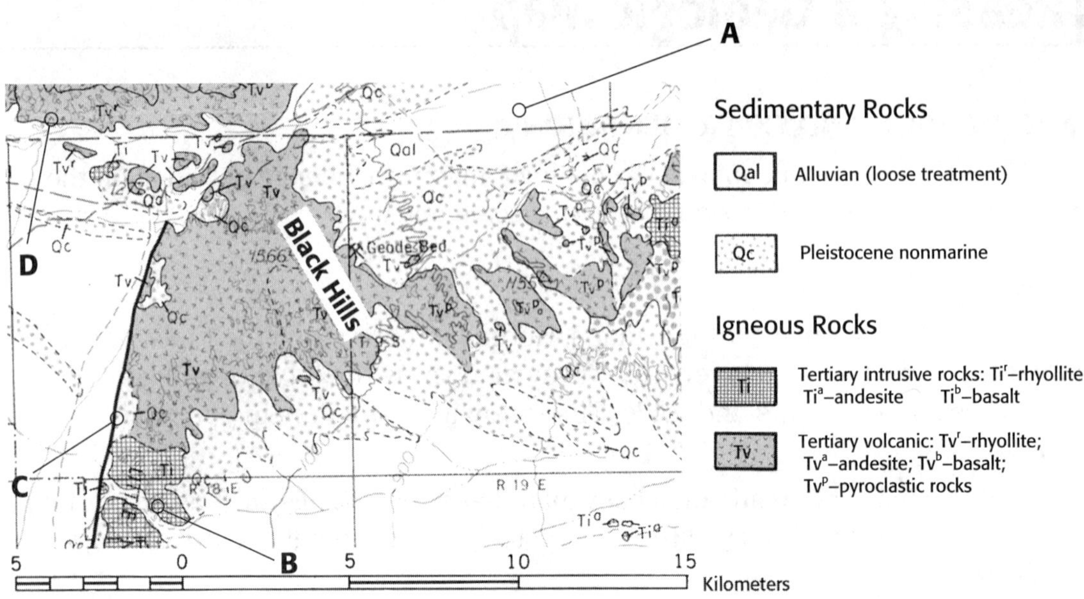

1. How many rock units are shown in this map? How many geologic time periods are represented?

____

____

2. Determine the age and the type of rock at point A and the age and the type of rock at point B.

____

____

3. What feature is represented by line C? What feature is represented by line D?

____

____

4. Find the Geode Bed on the map. If you wanted to find other areas rich in geodes, what type and age of rock would you look for?

____

____

Answer Key

Directed Reading A

SECTION: NATURAL RESOURCES

1. A
2. B
3. C
4. A
5. D
6. B
7. A
8. D
9. C
10. C
11. B

SECTION: ROCK AND MINERAL RESOURCES

1. B
2. A
3. C
4. B
5. C
6. A
7. A
8. D
9. D
10. D
11. D
12. A
13. B
14. D
15. C
16. A
17. A
18. C
19. B
20. D
21. B

SECTION: USING MATERIAL RESOURCES

1. A
2. B
3. A
4. D
5. B
6. C
7. C
8. petroleum
9. plastics

10. A
11. C
12. B
13. D
14. A
15. D
16. C
17. A
18. B
19. C

Directed Reading B

SECTION: NATURAL RESOURCES

1. They are Earth's resources or are made from Earth's resources.
2. C
3. D
4. the sun
5. a natural resource that can be replaced at the same rate at which the resource is used
6. Answers may vary. Sample answer: trees
7. a resource that forms at a much slower rate than the rate at which it is used
8. Answers may vary. Sample answer: coal, petroleum, natural gas
9. Answers may vary. Sample answer: People can turn off faucets when they are not in use.
10. Answers may vary. Sample answer: Polluted lakes and rivers can affect the water we drink.
11. Answers may vary. Sample answer: biking, walking, or taking a bus instead of driving a car; turning off lights when not in use
12. the process of reusing materials from waste or scrap
13. It takes less energy to recycle products than it does to make new ones.
14. Answers may vary. Sample answer: newspapers, aluminum cans, and cardboard boxes

SECTION: ROCK AND MINERAL RESOURCES

1. mineral
2. in a variety of environments in Earth's crust
3. rock
4. E
5. B
6. D
7. C
8. A
9. (1) Minerals are naturally formed. (2) Minerals are solids. (3) Minerals usually form by inorganic processes. (4) Minerals are crystals. (5) Minerals have consistent chemical compositions.
10. so they can be made into objects we need
11. ore
12. surface mining, subsurface mining
13. surface mines
14. strip mining
15. subsurface mine
16. tunnel
17. Mining can destroy plant and animal habitats. The waste from a mine can pollute water.
18. reclamation
19. diamond
20. Metals have shiny surfaces, are opaque to light, and are good conductors of heat and electricity.
21. Nonmetals have shiny or dull surfaces, may be translucent to light, and are good insulators of heat and electricity.
22. calcite
23. fertilizer, plaster board for construction, and plaster of Paris

SECTION: USING MATERIAL RESOURCES

1. Earth
2. B
3. A
4. atmosphere, crust, oceans, organisms
5. oxygen
6. F
7. E
8. D
9. A
10. B
11. C
12. petroleum

13. to obtain the stored energy in the plants
14. Answers may vary. Sample answer: fruit, paper, and maple syrup
15. Answers may vary. Sample answer: meat, leather, dairy products, fertilizers, and cooking fuel
16. Answers may vary. Sample answer should include two of the following: planting, feeding, fertilization, irrigation, pest control.
17. The total cost of making a product must be less than the price that a buyer is willing to pay for it.
18. a cheaper resource
19. Answers may vary. Sample answer: Habitats may be disrupted, or old growth forests may be destroyed.
20. reduced emission of pollutants, reclamation of mined land, replanting of harvested trees

Vocabulary and Section Summary A

SECTION: NATURAL RESOURCES

1. natural resource: any natural material that is used by humans, such as water, petroleum, minerals, forests, and animals
2. renewable resource: a natural resource that can be replaced at the same rate at which the resource is consumed
3. nonrenewable resource: a resource that forms at a rate that is much slower than the rate at which the resource is consumed
4. recycling: the process of recovering valuable or useful materials from waste or scrap

SECTION: ROCK AND MINERAL RESOURCES

1. mineral: a natural, usually inorganic solid that has a characteristic chemical composition and an orderly internal structure
2. ore: a natural material whose concentration of economically valuable minerals is high enough for the material to be mined profitably

SECTION: USING MATERIAL RESOURCES

1. material resource: a natural resource that humans use to make objects or to consume as food and drink
2. petroleum: a liquid mixture of complex hydrocarbon compounds; used widely as a fuel source

Vocabulary and Section Summary B

SECTION: NATURAL RESOURCES

1. natural resource
2. recycling
3. nonrenewable resource
4. renewable resource
5. temperature

SECTION: ROCK AND MINERAL RESOURCES

1. rock
2. reclamation
3. mineral
4. strip mining
5. ore
6. metamorphism
7. common objects

SECTION: USING MATERIAL RESOURCES

Across
2. plastics
3. material resource
4. recycling

Down
1. mineral
2. petroleum

Reinforcement

WHAT ARE MY RESOURCES?

1. R
2. N
3. R
4. R
5. N

Critical Thinking

1. Answers may vary. Sample answer: bed, rug, light, coffeepot, coffee, newspaper, eggs, bread, toaster, water, shower, toothpaste and toothbrush, clothes, car
2. Answers may vary. Sample answer: Yes, paper comes from trees; however, new forests need to be planted to replace the trees that are cut down.
3. Answers may vary. Sample answer: Material resources for making bread would include wheat or other grain, soil, and water, plus the ingredients to make the bread: flour, water, eggs, salt, and the metals that the oven is made of.
4. Answers may vary. Sample answer: The demand for these natural resources would be greater in an apartment building because more people live in it than in a single-family house.
5. Answers may vary. Sample answer: Sally could take the bus instead of driving her car. She could borrow a newspaper and recycle her clothes. She could also take short showers and not let the water run while brushing her teeth. She could also make only as much coffee as she plans to drink.

SciLinks Activity

Answers may vary. Students should include the manufacturing steps that are used in recycling the object they choose, as well as the recycled product that results. They should also explain the advantages of recycling in their chosen case.

Section Review

SECTION: NATURAL RESOURCES

1. Sample answer: A natural resource is any natural material that is used by humans. Renewable resources can be replaced at the same rate at which they are consumed. Nonrenewable resources form at a rate much slower than the rate at which they are consumed. Recycling is the process of recovering valuable materials from waste or scrap.

2. Humans use most natural resources by using products made from the resources.
3. Renewable resources form more quickly than nonrenewable resources form.
4. Sample answer: Three ways to conserve natural resources are: use resources only when necessary, reuse resources whenever possible, and recycle resources when possible.
5. Answers may vary. Students should note that most human activity affects Earth's resources.
6. 8.6 h × 3.3 L/h = 28.38 L
7. If a human population increase causes humans to use trees at a faster rate, the trees may be used at a rate faster than the rate at which they can grow back. If this happens, the trees might become a nonrenewable resource.

SECTION: ROCK AND MINERAL RESOURCES

1. Sample answer: A mineral is a naturally occurring, inorganic solid that has a definite crystal structure. Ore is a valuable material found in nature that has a high enough concentration to be mined profitably.
2. For a substance to be a mineral, it must be solid, be naturally occurring, have a crystalline structure, have a consistent chemical composition, and (usually) be inorganic.
3. Most minerals are nonrenewable resources.
4. The two main types of mining are surface mining and subsurface mining. In surface mining, material from Earth is extracted from the surface down. Subsurface mining is done by digging shafts or passageways to the area under Earth's surface from which the ore will be removed.
5. Sample answer: Gold can be made into jewelry, aluminum can be made into bike frames, and copper can be made into electrical wire.
6. Halite and gypsum might be found at the site of an ancient ocean.
7. Sample answer: Reclamation reduces the harmful effects of mining by

returning the land to its original state.
8. Recycling mineral products reduces the cost of manufacturing objects from minerals because the costs of mining the mineral are saved by using recycled mineral materials.
9. 0.007 × 120,000 metric tons = 840 metric tons
10. Sample answer: Two factors that could influence what concentration of a mineral is considered an ore are economic factors and technology. If the demand for a mineral increases, the price people are willing to pay for the mineral would go up. A deposit that was not considered an ore previously might be mined profitably at the new, higher price. (If the price goes down because of reduced demand, a deposit once considered an ore may no longer be an ore.) If a new technology is introduced, the new technology may reduce the cost of mining. Then, a deposit that was previously not an ore might be profitable to mine and be considered an ore.

SECTION: USING MATERIAL RESOURCES

1. Sample answer: A material resource is a natural resource that humans use to make something or as food or drink. Petroleum is a liquid mixture of hydrocarbons found in Earth.
2. Sample answer: Three of the resources that come from the atmosphere are argon, nitrogen, and oxygen. Argon is used in light bulbs, nitrogen is used to make fertilizer, and oxygen is used to burn fuels.
3. Petroleum is used as a material resource to make waxes, tar, and asphalt. Chemicals separated out of petroleum are used to make polymers.
4. Sample answer: Common objects that are made from plant resources include cloth, lumber, and rope.
5. Protecting the environment sometimes increases the price for an object in the store because the cost of manufacturing is increased by the cost of pollution controls or by the cost for reclamation or reforestation.

6. Students should give an opinion about considering environmental costs and then support their answer.

7. Sample answer: Petroleum is nonrenewable because it takes millions of years to form. Trees are renewable because they can grow as fast as they are used if they are used wisely. Water is renewable because the water cycle replenishes the water supply. Coal is nonrenewable because it takes millions of years to form. Aluminum is nonrenewable because it does not form as quickly as it is being used.

8. (400 million tons − 350 million tons) ÷ 350 million tons = 0.14; 0.14 × 100 = 14% increase

9. Sample answer: Energy resources are used at every step of paper production. These energy resources are probably mostly petroleum products used to transport materials, and petroleum or coal used to generate electricity. Trees, water, and other material resources are used to make paper. Recycled paper is made by using paper products that have been recycled instead of using trees. Some energy resources are used to collect and transport the recycled paper, and energy resources are used to make the recycled paper. I think it makes sense to recycle paper because trees are saved and the amount of energy used to make recycled paper is probably less than the amount used to make paper from trees.

Chapter Review

1. Sample answer: In this sentence, distribution means the relative arrangement of objects in space.

2. Sample answer: A natural resource is anything found in nature that people use. A material resource is a natural resource that people eat or drink or use to make things.

3. A mineral is a natural solid that has certain characteristics. An ore is a natural material that has an economic value if enough of the material is present in one place to be mined profitably.

4. A renewable resource forms as quickly as it is used, while a nonrenewable resource forms more slowly than the rate at which it is used.

5. B

6. A

7. C

8. D

9. B

10. D

11. Humans use natural resources to produce energy, for food, and to make products.

12. The five characteristics of a mineral are: it is usually inorganic, it is naturally occurring, it is solid, it is crystalline, and it has a consistent chemical composition.

13. Sample answer: Halite can form when a body of salt water evaporates. Quartz can form in a magma body called a *pluton* beneath Earth's surface.

14. Sample answer: Bike frames and electrical wire are made of metals. Glass and plaster are made from nonmetallic minerals.

15. Sample answer: Cotton cloth and maple syrup come from plant resources. Wool cloth and meat come from animal resources.

16. Surface mining is used to mine mineral deposits that are at or near Earth's surface. Subsurface mining is used to mine mineral deposits that are too deep within Earth to be surface mined.

17. Sample answer: Tar and plastic bottles are made from petroleum.

18. Students' answers should include these steps: cutting trees, making wood chips, making pulp, making paper from pulp, and making paper products.

19. An answer to this exercise can be found at the end of the Teacher Edition.

20. Sample answer: Conserving renewable natural resources is important so that the resources aren't used up faster than they can be replenished.

21. Sample answer: You can reduce your use of electrical energy and water. You can reuse products whenever possible. You can recycle things that cannot be reused.

22. Sample answer: yes; If humans begin using a resource faster than it can be replenished, a resource that was once renewable could become nonrenewable.

23. iron, zinc

24. They are material resources because they are used to make products.

25. Sample answer: If aluminum and copper were recycled, the amounts shown on the graph would go down as people would be returning metals they had used to the manufacturing cycle.

26. Sample answer: Metals such as copper and gold have shiny surfaces, are opaque to light, and conduct heat and electricity. Nonmetals such as calcite and quartz (silica) have either shiny or dull surfaces, may be translucent to light, and are insulators of heat and electricity.

27. Sample answer: A mining company would take into account what the land was like before the mine had been dug, what the elevation and shape of the surface was, what kind of soil had been there, and what types of plants and animals were native to the habitat.

28. The environmental costs of producing paper would be included in the cost of the paper at the store if the companies producing the paper had invested in equipment and processes to protect the environment and then included those costs in the price of the paper. The environmental costs may not be included in the store price if the producers had not invested in environmental protection and did not include those costs in the price of the paper.

29. Sample answer: Plan 1 would be a project to find another resource that could be used to make a replacement part for the computers. The effect on the price of computers would depend on the cost of the new resource relative to the cost of the old one.
Plan 2 would involve recycling the part from discarded computers and reusing the material. If the recycling wasn't too costly, the price of computers might be unaffected or may even decrease.

30. $13,620 + 1,140 + 2,260 + 7,040 = 24,060$ tons

31. $13,620 \times 0.564 = 7,682$ tons

32. $(13,620 \times 0.564) + (1,140 \times 0.14) + (2,260 \times 0.208) + (7,040 \times 0.504) = 11,860$ tons

33. Sample answer: As petroleum becomes more scarce, we will try to find alternative resources to supply both energy and the materials we make out of petroleum. If alternative energy sources are found, then petroleum may still be used to make plastics and other products. If we cannot find alternative energy resources, we may be forced to find alternative material resources.

Section Quizzes

SECTION: NATURAL RESOURCES

1. B
2. C
3. A
4. A
5. C
6. D
7. B

SECTION: ROCK AND MINERAL RESOURCES

1. B
2. A
3. C
4. D
5. D
6. A
7. C
8. B
9. D

SECTION: USING MATERIAL RESOURCES

1. B
2. D
3. A
4. C
5. B
6. C
7. A

Chapter Test A

1. C
2. D
3. C
4. B

5. A
6. A
7. D
8. A
9. B
10. C
11. A
12. A
13. D
14. C
15. B
16. B
17. C
18. A
19. C
20. A
21. B
22. mineral
23. ore
24. rock
25. reclamation

Chapter Test B

1. D
2. A
3. C
4. A
5. B
6. C
7. D
8. A
9. C
10. B
11. A
12. C
13. A
14. B
15. C
16. B
17. A
18. A
19. A
20. C
21. D
22. B
23. E
24. A
25. C

Chapter Test C

1. natural resource
2. renewable
3. recycling
4. nonrenewable
5. petroleum
6. C
7. A
8. D
9. D
10. A
11. B
12. Answers may vary. Sample answer: Humans can conserve resources by taking only what is needed. An example would be to turn off the faucet when water is not being used. We can take care of natural resources by keeping water sources free of pollution. We can also recycle or reuse items. An example would be to take plastics to a local recycling center.
13. Two methods of mining minerals are surface and subsurface. Surface mining is used when minerals are located at or near Earth's surface. Examples of this type of mining are open-pit mines, quarries, and strip mines. Subsurface mining is used when minerals are located deep within Earth's surface. Tunnels and vertical shafts have to be dug in this type of mining.
14. (1) Evaporation of salt water: When a body of salt water dries up, minerals are left behind. (2) Metamorphism: Changes in pressure, temperature, or chemical makeup that happen during this process cause different minerals to form. (3) Deposition: When surface water or groundwater carry dissolved materials into lakes and seas, minerals sink to the ocean or lake floor. (4) Reactions between hot water and rocks: Dissolved metals and other elements crystallize out of hot magma to form new minerals. (5) Cooling of plutons: When magma moves upward and then cools slowly, mineral crystals form.

15. Answers may vary. Sample answer: Mining can harm the environment because it can destroy or disturb plant and animal habitats. In addition, waste products from mines can seep into water sources, polluting both surface water and groundwater.

16. Answers may vary. Sample answer: Four economic costs could include the addition of high-purity clays and other ingredients, human labor and safety, materials used, and energy needed. Environmental costs that could ultimately affect the cost of paper include pollution-control measures and the replanting of harvested trees.

17. Answers may vary. Sample answer: Trees and water are two renewable resources that can get used up too quickly. The need for material resources such as lumber, paper, and fuel results in forests being cut down faster than they can regrow. Population growth has resulted in the use of more water. Droughts also can bring about water shortages.

18. a. atmosphere; **b.** oxygen; **c.** crust; **d.** minerals; **e.** oceans; **f.** salt; **g.** organisms; **h.** clothing

Performance-Based Assessment

1. Answers may vary. Sample answer: Data from geological core samples go into a computer. Software can make a drawing of the area, which tells miners where to find gold-bearing rock.

2. Answers may vary. Sample answer: Gold is encased in tons of rock. Blasting creates large open-pit mines.

3. Answers may vary. Sample answer: It is sent to a mill where it is crushed very fine. The larger particles are separated from the ground rock. The smallest particles are dissolved out of the ground rock with a cyanide solution, which leaves toxic waste water behind.

4. Answers may vary. Sample answer: Waste water pollutes the groundwater. Local habitats may be destroyed or disturbed.

5. Answers may vary. Sample answer: We can reduce the harmful effects of mining through a process called reclamation. Land is refilled, shafts are closed, waste water is treated, and plants are added to the land.

6. Answers may vary. Sample answer: Gold is used in jewelry, dental fillings, high-tech products, airplanes, cars, and telescopes. It is also used as money, representing wealth.

7. Answers may vary. Sample answer: Gold has many uses, but it is important to return the mined land to its original state.

Explore Activity

DATASHEET A

5. Answers may vary. Sample answer: A calculator is made of plastic parts and metal screws. Plastic is made from chemicals that are separated out of petroleum or crude oil. The metal ore for the screws is mined from rocks and minerals.

6. Sample answer: the costs of human labor and safety, the costs of raw materials and energy, and the costs of packaging, distributing, and selling the object

DATASHEET B

6. Sample answer: Costs included in the purchase price of an object are the cost of the raw materials, the cost of manufacturing the object (energy, labor), and the cost of packaging, distributing, and selling the object.

DATASHEET C

5. Answers may vary. Sample answer: A calculator is made of plastic parts and metal screws. Plastic is made from chemicals that are separated out of petroleum or crude oil. The metal ore for the screws is mined from rocks and minerals.

6. Sample answer: Costs included in the purchase price of an object are the cost of the raw materials, the cost of manufacturing the object (energy, labor), and the cost of packaging, distributing, and selling the object. One environmental cost might be a cost for reclamation of mined land.

Quick Lab: Renewable or Not?

DATASHEET A

4. higher
5. the use of a resource; the formation of a resource
6. A resource is renewable if it can be replaced at the same rate as the rate at which it is used. A resource is nonrenewable if it is replaced at a rate that is slower than the rate at which it is used.

DATASHEET B

4. The more quickly water is poured into the cup, the higher the water level is.
5. Water pouring out of the cup represents the use of a resource. Water pouring into the cup represents the formation of a resource.
6. A resource is renewable if it can be replaced at the same rate as the rate at which it is used. A resource is nonrenewable if it is replaced at a rate that is slower than the rate at which it is used.

DATASHEET C

4. The rate at which you pour the water into the cup controls the water level in the cup. The more quickly water is poured into the cup, the higher the water level is.
5. Water pouring out of the cup represents the use of a resource. Water poured into the cup represents the formation of a resource.
6. A resource is renewable if it can be replaced at the same rate as the rate at which it is used. A resource is nonrenewable if it is replaced at a rate that is slower than the rate at which it is used.

Quick Lab: Chocolate Ore

DATASHEET A

5. Answers may vary. Students should calculate the percentage as follows: (mass of chocolate ÷ mass of whole cookie) × 100 = % chocolate
6. Answers may vary. If different brands of cookies were used, students can compare the percentage of chocolate ore in the different brands tested.

7. Answers may vary. Students should recognize that if the percentage of chocolate in their cookie is greater than 0.7%, then the cookie is chocolate ore.

DATASHEET B

5. Answers may vary. Students should calculate the percentage as follows: (mass of chocolate ÷ mass of whole cookie) × 100 = % chocolate
6. Answers may vary. If different brands were used, students can compare the percentage of chocolate ore in the different brands tested.
7. A natural material is considered an ore if the percentage of the material to be mined is high enough that it can be mined profitably. Answers may vary. Students may say that the cookie was not ore because they could not sell the chocolate they "mined" for enough money to pay for their labor.

DATASHEET C

5. Answers may vary. Students should calculate the percentage as follows: (mass of chocolate ÷ mass of whole cookie) × 100 = % chocolate
6. Answers may vary. If different brands of cookies were used, students can compare the percentage of chocolate ore in the different brands tested.
7a. Answers may vary. A natural material is considered an ore if the percentage of the material to be mined is high enough that it can be mined profitably. Answers may vary. Students may say that the cookie was not ore because they could not sell the chocolate they "mined" for enough money to pay for their labor.
7b. Subsurface mining or strip mining may be chosen with a valid explanation.
7c. The amount of reclamation will be based on how carefully the students removed the chocolate. The chances of reclamation are probably low because the "mining" may have broken up the cookie.

Quick Lab: Products from Plants

DATASHEET A

3. Sawdust comes from trees, flour and liquid starch come from plants, and water comes from a lake or underground source. All of these resources are renewable if they are used wisely.

4. Sample answer: Trees are also used to make wood and paper; plants are used to make cloth and food products; and water is used to make food and pharmaceutical and personal care products.

DATASHEET B

3. Sawdust comes from trees, flour and liquid starch come from plants, and water comes from a lake or underground source. All of these resources are renewable if they are used wisely.

4. Sample answer: Trees are also used to make wood and paper. Plants are used to make cloth and food products. Water is used to make food, pharmaceutical, and personal care products.

DATASHEET C

3. Sawdust comes from trees, flour and liquid starch come from plants, and water comes from a lake or underground source. All of these resources are renewable if they are used wisely.

4. Sample answer: Trees are also used to make wood and paper. Plants are used to make cloth and food products. Water is used to make food and pharmaceutical and personal care products.

Chapter Lab

DATASHEET A

2. Answers may vary.

6. Answers may vary. Check to be sure that the correct equation was used and that the answers make sense.

7. Answers may vary. Check to be sure that the correct equation was used and that the answers make sense.

8. Answers may vary. Similarities may be due to similar lunches being purchased at the cafeteria. Differences may be due to different lunches being brought from home or to different products being purchased for lunch.

9. *paper and wood:* trees, renewable; *plastic:* petroleum, nonrenewable; *metal:* Earth (or minerals or ore), nonrenewable; *glass:* sand, nonrenewable (Note: while sand is nonrenewable because its rate of deposition is slow, sand is so abundant that it may be considered almost inexhaustible as a resource for the production of glass.)

10. Answers may vary. Check to see that students are adding the correct categories.

11. Answers may vary. Students should multiply the average mass per person by the total population of the school to determine the total amount of leftover lunch materials.

12. Sample answer: The school could set up a recycling program for glass and metal materials; the school could limit the use of nonrecyclable materials.

DATASHEET B

6. Answers may vary. Check to be sure that the correct equation was used and that the answers make sense.

7. Answers may vary. Check to be sure that the correct equation was used and that the answers make sense.

8. Answers may vary. Similarities may be due to similar lunches being purchased at the cafeteria. Differences may be due to different lunches being brought from home or to different products being purchased for lunch.

9. *paper and wood:* trees, renewable; *plastic:* petroleum, nonrenewable; *metal:* Earth (or minerals or ore), nonrenewable; *glass:* sand, nonrenewable (Note: while sand is nonrenewable because its rate of deposition is slow, sand is so abundant that it may be considered almost inexhaustible as a resource for the production of glass.)

10. Answers may vary. Check to see that students are adding the correct categories.

11. Answers may vary. Students should explain that they can multiply the average mass per person by the total population of the school to determine the total amount of leftover lunch materials.

12. Sample answer: The school could set up a recycling program for glass and metal materials.

DATASHEET C

2. Answers may vary. Sample answer: When we separate the materials left over from lunch, we will find 40% are paper and wood products, 40% are plastic products, 10% are materials made of metal, and 10% are materials made of glass. Since paper and plastic products make up most of the food packaging at our school, they will have the largest percentages.

6. Answers may vary. Check to be sure that the correct equation was used and that the answers make sense.

7. Answers may vary. Check to be sure that the correct equation was used and that the answers make sense.

8. Answers may vary. Similarities may be due to similar lunches being purchased at the cafeteria. Differences may be due to different lunches being brought from home or to different products being purchased for lunch.

9. *paper and wood:* trees, renewable; *plastic:* petroleum, nonrenewable; *metal:* Earth (or minerals or ore), nonrenewable; *glass:* sand, nonrenewable (Note: while sand is nonrenewable because its rate of deposition is slow, sand is so abundant that it may be considered almost inexhaustible as a resource for the production of glass.)

10. Answers may vary. Check to see that students are adding the correct categories.

11. Answers may vary. Students should multiply the average mass per person by the total population of the school to determine the total amount of leftover lunch materials in each category. Then, multiply the total in each category by the number of days in the school year.

12. Sample answer: The school could set up a recycling program for glass and metal materials; the school could limit the use of nonrecyclable materials. To carry out the recycling program, the school could put up posters explaining the program and place labeled bins in the cafeteria and other locations. To limit the use of nonrecyclable materials, a committee of students and teachers could do research to find and purchase recyclable substitutes for these materials.

Science Skills Activity

DATASHEET

1. Four different rock units are shown in this map. Two different periods are represented, the Quaternary and Tertiary Periods.

2. At point A there is alluvium (loose sediment) from the Quaternary Period. At point B there is intrusive igneous rock from the Tertiary Period.

3. Line C represents a fault. Line D represents a contact between two rock units.

4. Sample answer: It looks like the Geode Bed is near the contact of the Pleistocene sedimentary rock and the Tertiary volcanic rock. I would try to determine whether the geodes were found mostly at the contact between the two types of rock or in one or the other type, and then look in similar areas for other areas rich in geodes.